Introduction to Digital Filters

Other Macmillan titles of related interest

Macmillan New Electronics Series

Introduction to Digital Filters

Trevor J. Terrell

Principal Lecturer,
Lancashire Polytechnic

Second Edition

MACMILLAN
EDUCATION

First edition 1980
Reprinted 1983, 1984, 1985
Second edition 1988

Published by
MACMILLAN EDUCATION LTD
Houndmills, Basingstoke, Hampshire RG21 2XS
and London
Companies and representatives
throughout the world

Printed in Hong Kong

British Library Cataloguing in Publication Data
Terrell, Trevor J.
    Introduction to digital filters.—2nd ed.
    1. Digital filters (Mathematics).
    I. Title
    621.3815'324   QA297

    ISBN 0–333–44322–5
    ISBN 0–333–44323–3 Pbk

To Jennifer, Janet and Lesley

Contents

Preface to the First Edition

In the 1970s computer technology has progressed at a phenomenal rate and, as microprocessor systems continue to develop, the cost of implementing digital signal processors will, hopefully, correspondingly decrease. Consequently in the 1980s it is expected that the number of practical applications of digital filters will significantly increase; furthermore it is possible that many established continuous-time filter systems will be replaced by equivalent digital filter systems. There is therefore a growing need for more well-trained engineering and science graduates, postgraduates and practising engineers who are skilled in the design and implementation of digital filter systems. Fortunately many polytechnics and universities have already recognised the importance of digital signal processing, and this subject now generally forms part of the core material in many undergraduate and postgraduate engineering courses.

This book presents a concise introduction to the fundamental techniques involved in the design and implementation of digital filters. The level of the information presented makes it suitable for use mainly in the second and the final years of electrical and electronic engineering undergraduate courses. Some of the material goes beyond the undergraduate level and will be useful to postgraduate students and practising engineers. The book includes many appropriate worked examples which serve to illustrate theoretical concepts as they are developed in the text. Indeed, the main purpose of the worked examples is to help to bridge the gap between the theoretical and practical aspects of digital filters, and it is intended that this feature will greatly assist the reader (especially the undergraduate student) in relating the theory to practical results.

The first chapter introduces sampled-data signals and systems, and the basic required mathematical concepts of the Z-transform and the inverse Z-transform are developed. In chapter 2 a number of useful design methods applicable to recursive digital filters are described in detail. The design methods applicable to non-recursive digital filters are examined in chapter 3. The main considerations that must be given to quantisation effects in the practical implementation of digital filters are discussed in chapter 4. In chapter 5 (the final chapter) hardware and software aspects of digital filter implementation are discussed in detail, and some pertinent microprocessor system concepts are described. The appendixes at the end of the book have been included to help the reader who wishes to undertake the task of translating a 'paper design' into a practical working digital filter. Typically the latter may be a project for the undergraduate student, or indeed in some cases a postgraduate student, or project development undertaken by the practising engineer.

I wish to express my thanks to my colleague Dr R. J. Simpson (Preston Polytechnic) for his encouragement and helpful advice during the preparation and writing of this book. I also wish to express my thanks to Mr G. Collins (Preston Polytechnic) for his invaluable assistance in sorting out practical problems associated with software used in implementing some of the digital filter designs presented in the book. I gratefully thank Dr E. T. Powner (UMIST) and Dr M. G. Hartley (UMIST) for giving me the opportunity to teach some of the material in this book to postgraduate students on their M.Sc. Digital Electronics course. My thanks also go to Mrs T. Smith for her cheerfulness and competence in typing the manuscript. I especially thank my wife, Jennifer, for her love, patience, encouragement and understanding.

TREVOR J. TERRELL

Preface to the Second Edition

In the years since this book was first published (1980) there has been a remarkable growth in the development of DSP devices and their applications. The importance of digital filters in system implementations is now well known, and the forecast made in the Preface to the First Edition that we would witness a growing need for more well-trained engineers, skilled in the design of digital filters, has been substantiated by the demand shown by industry. It is intended that the second edition of this book will continue to provide a firm base-of-knowledge of the principles of digital filter design and implementation, and it is aimed at students and engineers working with digital filters. It is hoped, therefore, that this book will continue to meet the needs of industry and academia, and the precept of presenting a concise introduction to the fundamental techniques of digital filter design and implementation has been preserved in this second edition.

The section on the Fast Fourier Transform in chapter 1 has been expanded, and because of the growth in image processing applications, the two-dimensional FFT is described. Chapter 5 has been extensively revised, the main changes being (i) the inclusion of a complete microprocessor-based design example with solution and (ii) details of the TMS320C10 DSP device, with a case study of its application to the companding of speech signals. These changes reflect the developments in the techniques for implementing digital filters, and they illustrate how microprocessors may be used to achieve the desired processing operations. Also minor changes and corrections have been undertaken to improve the book, and some of the more dated material has been removed.

In revising the book a number of people were consulted and a number of sources of reference were used, and I wish to express my sincere thanks to everyone who assisted me. In particular, I am grateful to my colleague Mike Peak for his help with the FORTRAN 77 program for the FFT implementation and the work of Zahid Hafeez in producing the results shown in figure 1.20 is gratefully acknowledged.

TREVOR J. TERRELL

Acknowledgements

The author and publisher acknowledge the permission of Texas Instruments to use copyright material for the TMS320C10 device, which is reproduced in chapter 5.

The permission of Butterworth Scientific Ltd to use the appendix listing in chapter 5, originally published in *Microprocessors and Microsystems*, is also gratefully acknowledged.

1 Introduction

1.1 BACKGROUND TO DIGITAL FILTERS

In the decade 1960–1970 high speed digital computers developed rapidly and they became widely available for processing the digital representation of electrical waveforms. Consequently it became possible to use the basic established theoretical concepts of Fourier analysis, waveform sampling, Z-transforms, etc., in digital filter design. Hence digital filtering of electrical signals became a practical reality as computer technology developed.

In 1965 the Cooley-Tukey algorithm[1] was published, thereby making an important contribution to the development of digital signal processing. Another notable contribution to the early development of digital filters was made by James Kaiser[2], when he worked on the design of filters using the bilinear Z-transform. However, these were not the only significant contributions, and from the mid 1960s onwards many contributions to research and development of digital filters have been published.[3,4]

Digital signal processing has become an established method of filtering electrical waveforms, and the associated theory of discrete-time systems can often be employed in a number of science and technology disciplines.[5] There are many applications of digital signal processing, typically these are: analysis of biomedical signals; vibration analysis; picture processing; analysis of seismic signals and speech analysis.

Since the 1970s computer technology has progressed at a phenomenal rate. The existence of relatively inexpensive LSI digital circuits, microcomputers, and A/D and D/A converters means that digital filtering has become widely used, and it is quite possible that many established continuous-time filter systems will be replaced by equivalent digital filter systems. As microprocessor and VLSI integrated-circuit technology continues to develop the cost of implementing digital filters will probably decrease, and we can foresee that there will be a corresponding increase in the number of practical applications. Credence is added to this prediction when one takes into account the inherent advantages of digital filters, namely

1

(1) they do not drift;

(2) they can handle low frequency signals;

(3) frequency response characteristics can be made to approximate closely to the ideal;

(4) they can be made to have no insertion loss;

(5) linear phase characteristics are possible;

(6) adaptive filtering is relatively simple to achieve;

(7) digital word-length may be controlled by the filter designer, and therefore the accuracy of the filter may be precisely controlled; and

(8) cost and availability of hardware is not generally a problem.

A typical digital filtering process is shown schematically in figure 1.1. Referring to figure 1.1, the input signal, $x(t)$, is sampled regularly at instants T seconds apart, and each sample is converted to a digital-word, thus forming the digital input sequence $x^*(t)$. The digital filter, $G(Z)$, operates on the sequence $x^*(t)$ to form the output sequence $y^*(t)$. The sequence $y^*(t)$ is converted to a train of pulses, $f(t)$, the area of each pulse being equal to T times the respective sequence

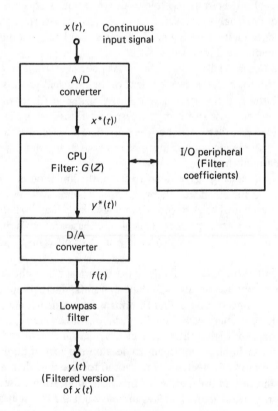

Figure 1.1 Block diagram of a typical digital filtering process

value. To recover the continuous output, $y(t)$, the train of pulses, $f(t)$, may be passed through a simple analogue R-C lowpass filter. Clearly then, the sampled-data signals $x^*(t)$ and $y^*(t)$ play an important part in the digital filtering process; consequently this type of signal will be discussed in detail in the following section.

1.2 SAMPLED-DATA SIGNALS[6,7]

A simple, but adequate, model of the sampling process is one which considers the continuous input signal, $x(t)$, to be sampled by a switch closing periodically for a short time, τ seconds, with a sampling interval of T seconds (see figure 1.2). Referring to figure 1.2 it is seen that the switch output is a train of finite width pulses.

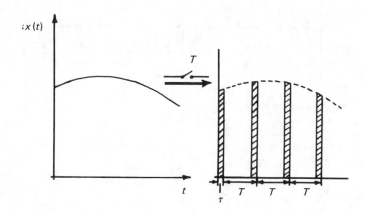

Figure 1.2 Model of the sampling process

However, if the pulse width, τ, is negligible compared with the interval between successive samples, T, the output of the sampler can be considered to be a train of *impulses* with their heights proportional to the values of $x(t)$ at the sampling instants (see figure 1.3).

The ideal sampling function, $\delta_T(t)$, represents a train of unit-impulses, and it is defined as

$$\delta_T(t) = \sum_{n=-\infty}^{\infty} \delta(t - nT) \tag{1.1}$$

where $\delta(t)$ is the unit-impulse function occurring at $t = 0$, and $\delta(t - nT)$ is a delayed impulse function occurring at $t = nT$. Therefore

$$x^*(t) = x(t) \sum_{n=-\infty}^{\infty} \delta(t - nT) \tag{1.2}$$

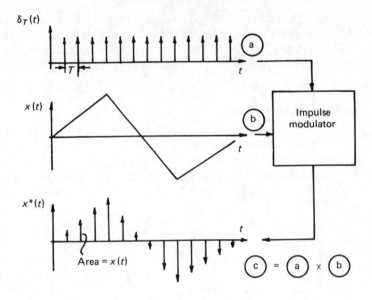

Figure 1.3 Impulse sampling process

but the value of $x(t)$ is only known for $t = nT$, and furthermore for a physical system $x(t) = 0$ for $t < 0$, therefore

$$x^*(t) = \sum_{n=0}^{\infty} x(n)T\,\delta(t - nT) \qquad (1.3)$$

Thus we see that $x^*(t)$ is a weighted sum of shifted unit-impulses (see figure 1.3, waveform c).

Referring to equation 1.1 we can expand $\delta_T(t)$ as a Fourier series, that is

$$\delta_T(t) = \sum_{n=-\infty}^{\infty} C_n\, e^{\,jn\omega_s t}$$

where

$$C_n = 1/T \int_0^T \delta_T(t)\, e^{-jn\omega_s t}\, dt$$

and ω_s is the sampling frequency equal to $2\pi/T$ rad/s. Since the area of an impulse is unity, then

$$\int_0^T \delta_T(t)\, e^{-jn\omega_s t}\, dt = 1$$

and therefore $C_n = 1/T$, hence

$$\delta_T(t) = 1/T \sum_{n=-\infty}^{\infty} e^{jn\omega_s t}$$

and we have seen in figure 1.3 that for the impulse modulator $x^*(t) = \delta_T(t)x(t)$, therefore

$$x^*(t) = \frac{1}{T} \sum_{n=-\infty}^{\infty} x(t) e^{jn\omega_s t} \tag{1.4}$$

Now taking Laplace transforms and using the associated shifting theorem we obtain

$$X^*(S) = \mathcal{L}[x^*(t)] = \frac{1}{T} \sum_{n=-\infty}^{\infty} X(S - jn\omega_s)$$

therefore

$$X^*(j\omega) = \frac{1}{T} \sum_{n=-\infty}^{\infty} X[j(\omega - n\omega_s)] \tag{1.5}$$

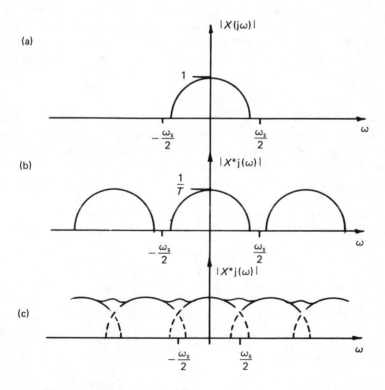

Figure 1.4 (a) Frequency spectrum of the continuous signal; (b) frequency spectra of the sampled signal; (c) aliasing of frequency spectra

Thus we see from equation 1.5 that as a result of impulse-sampling the frequency spectrum of $x(t)$ is repeated ad infinitum at intervals of $j\omega_s$. Let us now consider the frequency spectra of $x^*(t)$ (see figure 1.4). Referring to figure 1.4, if $\omega_s/2$ is greater than the highest frequency component of $x(t)$ (see figure 1.4a), then the original signal can theoretically be recovered from the spectra of $x^*(t)$ (see figure 1.4b). In contrast if $\omega_s/2$ is not greater than the highest frequency component in the continuous signal (see figure 1.4c), then *folding* of the frequency response function occurs, and consequently the original signal cannot be reclaimed from the sampled-data signal. The errors caused by folding of the frequency spectra are generally referred to as *aliasing errors*, which may be avoided by increasing the sampling frequency, provided the filter's physical constraints permit this to occur. Also it is important to note that the baseband spectra of $x^*(t)$ are amplitude scaled by a factor equal to $1/T$ compared with the spectrum of $x(t)$ (see equation 1.5).

It has been established that the sampled-data signal has an infinite number of complementary frequency spectra, which means that there must be an infinite number of associated pole–zero patterns in its S-plane representation. Consequently the analysis of any sampled-data signal or system is extremely difficult when working in the S-plane. However, fortunately it is possible to *Z-transform* the sampled signal or system, thereby yielding a mathematical description that is relatively simple to analyse. The Z-transform will be discussed in detail in the following section.

1.3 THE Z-TRANSFORM[8]

This transformation is used to describe the properties of a sampled-data signal or system, and as we shall see later it provides useful methods of representing the sampled-data signal or system by either a finite set of *poles* and *zeros* (frequency-domain representation) or by a *linear difference equation* (time-domain representation).

The Z-transform is simply a rule that converts a sequence of numbers into a function of the complex variable Z, and it has inherent properties that enable

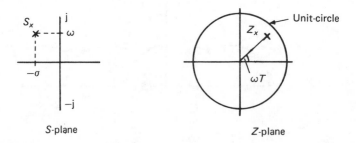

Figure 1.5 S-plane to Z-plane transformation

linear difference equations to be solved using straightforward algebraic manipulations. However, we will postpone any further discussion of the practical application of the Z-transform until the mathematical concepts of this topic have been discussed and demonstrated. We will now consider the basic mathematical description of the Z-transform and some of its important properties.

Suppose that we let $Z = e^{ST} = e^{-(\sigma - j\omega)T}$, then $|Z| = e^{\sigma T}$ and $\underline{/Z} = \omega T$, so that any point S_x in the S-plane transforms to a corresponding point Z_x in the Z-plane, as shown in figure 1.5. Now referring to table 1.1 it is seen that the imaginary axis in the S-plane transforms (maps) to the circumference of the unit-circle in the Z-plane. When σ is negative $|Z| < 1$ and when σ is positive $|Z| > 1$. Hence a strip ω_s wide in the left-hand half of the S-plane transforms to the area inside the unit-circle in the Z-plane, and this same strip in the right-hand half of the S-plane transforms to the area outside the unit-circle in the Z-plane (see figure 1.6).

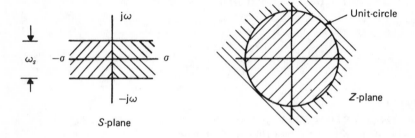

Figure 1.6 Relation between S-plane and Z-plane regions

Table 1.1

$\sigma = 0,\ \omega_s = 2\pi/T$	
$j\omega$	$Z = 1\ \underline{/\omega T}$
0	$1\ \underline{/0^\circ}$
$\omega_s/8$	$1\ \underline{/45^\circ}$
$\omega_s/4$	$1\ \underline{/90^\circ}$
$3\omega_s/8$	$1\ \underline{/135^\circ}$
$\omega_s/2$	$1\ \underline{/180^\circ}$
$5\omega_s/8$	$1\ \underline{/225^\circ}$
$3\omega_s/4$	$1\ \underline{/270^\circ}$
$7\omega_s/8$	$1\ \underline{/315^\circ}$
ω_s	$1\ \underline{/360^\circ}$

The most important effect of the Z-transformation is that since the poles and zeros of $x^*(t)$ are spaced along the imaginary axis at intervals of $\omega_s = 2\pi/T$ rad/s, all sets of poles and zeros in the S-plane transform to a single set of poles and zeros in the Z-plane—strictly the sets of poles and zeros are superimposed on the Z-plane lying in Riemann surfaces, and they are treated effectively as a single set of poles and zeros.

Now referring to equation 1.3 we see that

$$x^*(t) = x(0)T\delta(t) + x(1)T\delta(t - T) + x(2)T\delta(t - 2T) + \ldots$$

and taking Laplace transforms we obtain

$$\mathcal{L}[x^*(t)] = x(0)T + x(1)T e^{-ST} + x(2)T e^{-2ST} + \ldots$$

but for the Z-transform we have $Z = e^{ST}$, that is $Z^{-1} = e^{-ST}$, and therefore $Z^{-n} = e^{-nST}$. The term Z^{-n} represents a time delay of n sampling periods. Thus the Z-transform of $x^*(t)$ is

$$X(Z) = x(0)T + x(1)T Z^{-1} + x(2)T Z^{-2} + \ldots$$

that is

$$X(Z) = \sum_{n=0}^{\infty} x(n)T Z^{-n} \tag{1.6}$$

Example 1.1
An input sequence corresponding to a sampled-data signal is $\{1, 0.5, 0, -0.8, -3.\}$ What is the Z-transform of the input sequence?

SOLUTION
Using equation 1.6 we obtain $X(Z) = 1 + 0.5Z^{-1} - 0.8Z^{-3} - 3Z^{-4}$.

Example 1.2
Suppose that the input signal of a digital filter is $x(t) = e^{-at}$; what is the Z-transform of $x^*(t)$?

SOLUTION
$x^*(t) = \{e^{-0}, e^{-aT}, e^{-2aT}, e^{-3aT}, \ldots\}$, therefore using equation 1.6 we obtain $X(Z) = 1 + e^{-aT}Z^{-1} + e^{-2aT}Z^{-2} + e^{-3aT}Z^{-3} + \ldots$, and for $|Z| > e^{-aT}$, $X(Z)$ can be expressed in *closed form* as $X(Z) = Z/(Z - e^{-aT})$.

Example 1.2 is a particular case of a geometric sequence, and in general the Z-transform of the geometric sequence, g^n, is expressed as

$$Z(g^n) = \sum_{n=0}^{\infty} g^n Z^{-n}$$

(see equation 1.6) therefore

$$X(Z) = \begin{cases} Z/(Z-g) & \text{for } |Z| > |g| \\ \text{unbounded} & \text{for } |Z| < |g| \end{cases} \tag{1.7}$$

The set of Z in the Z-plane for which the magnitude of $X(Z)$ is finite is called the *region of convergence* (for example $|Z| > |g|$ in equation 1.7), and in contrast the set of Z in the Z-plane for which the magnitude of $X(Z)$ is infinite is called the *region of divergence* (for example $|Z| < |g|$ in equation 1.7). Figure 1.7 illustrates the regions of convergence and divergence for $\mathbf{Z}(g^n)$.

Example 1.3
Suppose that the input signal of a digital filter is $x(t) = \sin \omega t$; what is the Z-transform of $x^*(t)$?

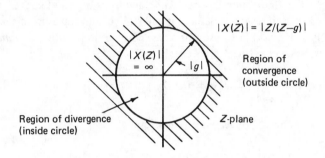

Figure 1.7 Z-plane regions of convergence and divergence for $\mathbf{Z}(g^n)$

SOLUTION
$x^*(t) = \sin n\omega T = (e^{jn\omega T} - e^{-jn\omega T})/j2$, therefore using equation 1.6 we obtain

$$X(Z) = \sum_{n=0}^{\infty} \left[\frac{(e^{jn\omega T} - e^{-jn\omega T})}{j2} \right] Z^{-n}$$

$$= (1/j2) \left[\sum_{n=0}^{\infty} (e^{jn\omega T}) Z^{-n} - \sum_{n=0}^{\infty} (e^{-jn\omega T}) Z^{-n} \right]$$

Now letting $e^{jn\omega T} = g^n = (e^{j\omega T})^n$, that is $g = e^{j\omega T}$, and using equation 1.7 we obtain

$$\sum_{n=0}^{\infty} (e^{jn\omega T}) Z^{-n} = \frac{Z}{(Z - e^{j\omega T})} \quad \text{for } |Z| > 1$$

Similarly letting $e^{-jn\omega T} = g^n = (e^{-j\omega T})^n$, then using equation 1.7 we obtain

$$\sum_{n=0}^{\infty} (e^{-jn\omega T}) Z^{-n} = \frac{Z}{(Z - e^{-j\omega T})} \quad \text{for } |Z| > 1$$

therefore

$$X(Z) = \frac{Z}{j2} \left[\frac{1}{Z - e^{j\omega T}} - \frac{1}{Z - e^{-j\omega T}} \right] \quad \text{for} \quad |Z| > 1$$

$$= \frac{Z}{j2} \left[\frac{e^{j\omega T} - e^{-j\omega T}}{Z^2 - (e^{j\omega T} + e^{-j\omega T})Z + 1} \right]$$

that is

$$X(Z) = \frac{(Z \sin \omega T)}{(Z^2 - 2Z \cos \omega T + 1)}$$

Equation 1.6 defines a power series, $X(Z)$, and in its region of convergence it may be differentiated several times and it will still be convergent in this region. Consequently it is possible to use this knowledge to derive other *Z-transform pairs*. That is, using equation 1.6 we obtain

$$\frac{d}{dZ} [X(Z)] = \sum_{n=0}^{\infty} -nx(n)TZ^{-(n+1)} \quad \text{for} \quad |Z| > R$$

where R is the radius of convergence. Therefore

$$-Z \left\{ \frac{d}{dZ} [X(Z)] \right\} = \sum_{n=0}^{\infty} nx(n)T Z^{-n}$$

that is

$$-Z \left\{ \frac{d}{dZ} [X(Z)] \right\} = \mathbf{Z}[(nx(n)T] \tag{1.8}$$

and

$$-ZT \left\{ \frac{d}{dZ} [X(Z)] \right\} = \mathbf{Z}[nTx(n)T] \tag{1.9}$$

Example 1.4
Suppose that $x(n)T$ is a geometric sequence, g^n, determine the Z-transform of (ng^n).

SOLUTION
From equation 1.7 we know that for the geometric sequence, g^n,
$X(Z) = Z/(Z - g)$ for $|Z| > |g|$. Therefore using equation 1.8 we obtain

$$\mathbf{Z}[ng^n] = -Z \left\{ \frac{d}{dZ} \left[\frac{Z}{Z - g} \right] \right\} = \frac{(Zg)}{(Z - g)^2}$$

Example 1.5
Suppose that the input signal of a digital filter is $x(t) = t\,e^{-at}$, determine the
Z-transform of $x^*(t)$.

SOLUTION
$x^*(t) = nT\,e^{-anT}$, and from example 1.2 we know that $X(Z) = Z/(Z - e^{-aT})$,
therefore using equation 1.9 we obtain

$$Z(nT\,e^{-anT}) = -ZT\left\{\frac{d}{dZ}\left[\frac{Z}{Z - e^{-aT}}\right]\right\} = \frac{(ZT\,e^{-aT})}{(Z - e^{-aT})^2} \quad \text{for} \quad |Z| > e^{-aT}$$

The Z-transform has a number of useful properties which can be used in our
study of digital filters. These properties will now be discussed.

1.3.1 Z-transform Properties

(a) Linearity Property
Consider $x(n)T$ expressed in the form $Ax_1(n)T + Bx_2(n)T$, where A and B are
constants, then using equation 1.6 we obtain

$$X(Z) = \sum_{n=0}^{\infty} (Ax_1(n)T + Bx_2(n)T)Z^{-n}$$

$$= A\sum_{n=0}^{\infty} x_1(n)TZ^{-n} + B\sum_{n=0}^{\infty} x_2(n)TZ^{-n}$$

that is

$$X(Z) = AX_1(Z) + BX_2(Z) \quad \text{for} \quad |Z| > \max(R_1, R_2) \tag{1.10}$$

where R_1 and R_2 are the radius of convergence for $X_1(Z)$ and $X_2(Z)$ respectively,
and the term $\max(R_1, R_2)$ is used to specify the larger of the two numbers R_1
and R_2.

Example 1.6
Suppose that the input signal of a digital filter is $x(t) = 3 + \sin \omega t$; determine the
Z-transform of $x^*(t)$

SOLUTION
$x(n)T = 3 + \sin n\omega T$, and $X(Z) = Z[x(n)T]$, therefore using equation 1.10 we
obtain

$$X(Z) = 3Z(1) + 1Z(\sin n\omega T)$$

but $Z(1) = Z/(Z - 1)$ (see table 1.2), and $Z(\sin n\omega T)$ has been derived in
example 1.3, therefore

$$X(Z) = \left\{3\left[\frac{Z}{Z - 1}\right] + 1\left[\frac{Z \sin \omega T}{Z^2 - 2Z \cos \omega T + 1}\right]\right\} \text{for } |Z| > 1$$

therefore

$$X(Z) = \frac{3Z^3 + (\sin \omega T - 6 \cos \omega T)Z^2 + (3 - \sin \omega T)Z}{Z^3 - (2 \cos \omega T + 1)Z^2 + (2 \cos \omega T + 1)Z - 1}$$

Table 1.2 A selection of some common Z-transforms

$f(n), n \geqslant 0$ or $f(t), t \geqslant 0$	$F(S)$	$F(Z) = \sum\limits_{n=0}^{\infty} f(n)\, Z^{-n}$ or $\sum\limits_{n=0}^{\infty} f(n)TZ^{-n}$
1	$\dfrac{1}{S}$	$\dfrac{Z}{Z-1}$
t	$\dfrac{1}{S^2}$	$\dfrac{TZ}{(Z-1)^2}$
t^2	$\dfrac{2}{S^3}$	$\dfrac{T^2 Z(Z+1)}{(Z-1)^3}$
t^3	$\dfrac{3!}{S^4}$	$\dfrac{T^3 Z(Z^2 + 4Z + 1)}{(Z-1)^4}$
t^4	$\dfrac{4!}{S^5}$	$\dfrac{T^4 Z(Z^3 + 11Z^2 + 11Z + 1)}{(Z-1)^5}$
e^{-at}	$\dfrac{1}{S+a}$	$\dfrac{Z}{Z - e^{-aT}}$
$t\,e^{-at}$	$\dfrac{1}{(S+a)^2}$	$\dfrac{TZ\,e^{-aT}}{(Z - e^{-aT})^2}$
$t^2\,e^{-at}$	$\dfrac{2}{(S+a)^3}$	$\dfrac{T^2\,e^{-aT}Z\,(Z + e^{-aT})}{(Z - e^{-aT})^3}$
$1 - e^{-at}$	$\dfrac{a}{S(S+a)}$	$\dfrac{Z(1 - e^{-aT})}{Z^2 - Z(1 + e^{-aT}) + e^{-aT}}$
$e^{-at} - e^{-bt}$	$\dfrac{b-a}{(S+a)(S+b)}$	$\dfrac{Z(e^{-aT} - e^{-bT})}{Z^2 - Z(e^{-aT} + e^{-bT}) + e^{-(a+b)T}}$
$\cos \omega t$	$\dfrac{S}{S^2 + \omega^2}$	$\dfrac{Z(Z - \cos \omega T)}{Z^2 - 2Z \cos \omega T + 1}$
$\sin \omega t$	$\dfrac{\omega}{S^2 + \omega^2}$	$\dfrac{Z \sin \omega T}{Z^2 - 2Z \cos \omega T + 1}$

$e^{-at} \cos \omega t$	$\dfrac{S + a}{(S + a)^2 + \omega^2}$	$\dfrac{Z^2 - Z\,e^{-aT} \cos \omega T}{Z^2 - 2Z\,e^{-aT} \cos \omega T + e^{-2aT}}$
$e^{-at} \sin \omega t$	$\dfrac{\omega}{(S + a)^2 + \omega^2}$	$\dfrac{Z\,e^{-aT} \sin \omega T}{Z^2 - 2Z\,e^{-aT} \cos \omega T + e^{-2aT}}$
$\cosh \omega t$	$\dfrac{S}{S^2 - \omega^2}$	$\dfrac{Z^2 - Z \cosh \omega T}{Z^2 - 2Z \cosh \omega T + 1}$
$\sinh \omega t$	$\dfrac{\omega}{S^2 - \omega^2}$	$\dfrac{Z \sinh \omega T}{Z^2 - 2Z \cosh \omega T + 1}$
a^n		$\dfrac{Z}{Z - a}$
n		$\dfrac{Z}{(Z - 1)^2}$
n^2		$\dfrac{Z(Z + 1)}{(Z - 1)^3}$
n^3		$\dfrac{Z(Z^2 + 4Z + 1)}{(Z - 1)^4}$
$\dfrac{a^n}{n!}$		$e^{a/Z}$
na^n		$\dfrac{aZ}{(Z - a)^2}$
$n^2 a^n$		$\dfrac{aZ(Z + a)}{(Z - a)^3}$
$\delta(n)T$		1
$\delta(n - m)T$		Z^{-m}

(b) Right-shifting Property

Consider a signal of the form $y(n)T = x(n - k)T$, which is 0 for $t < 0$, then using equation 1.6 we obtain

$$Y(Z) = \sum_{n=0}^{\infty} x(n - k)TZ^{-n}$$

$$= [x(-k)T + x(1 - k)TZ^{-1} + x(2 - k)TZ^{-2} + \ldots$$
$$+ x(-1)TZ^{-k+1} + x(0)TZ^{-k} + x(1)TZ^{-(k+1)} + \ldots]$$

but since $y(n)T = 0$ for $t < 0$, then

$$Y(Z) = [x(0)T + x(1)TZ^{-1} + x(2)TZ^{-2} + \ldots] Z^{-k}$$

therefore

$$Y(Z) = \left[\sum_{n=0}^{\infty} x(n)TZ^{-n} \right] Z^{-k}$$

therefore

$$Y(Z) = X(Z) Z^{-k} \quad \text{for} \quad |Z| > R \qquad (1.11)$$

where R is the radius of convergence of $X(Z)$.

Example 1.7

Suppose that a signal is defined by $y(n)T = 2x(n)T + 4y(n-2)T$; determine the
Z-transform of $y(n)T$.

SOLUTION

Taking $y(n)T = 0$ for $t < 0$ we have

$$y(n)T = 2x(n)T + 4y(n-2)T$$

Therefore

$$Y(Z) = 2X(Z) + 4Y(Z)Z^{-2}$$

and therefore

$$Y(Z) = \frac{2X(Z)}{(1 - 4Z^{-2})}$$

(c) Left-shifting Property

Consider a signal of the form $y(n)T = x(n+k)T$, which is 0 for $t < 0$, then using
equation 1.6 we obtain

$$Y(Z) = \sum_{n=0}^{\infty} x(n+k)TZ^{-n}$$

$$= x(k)T + x(1+k)TZ^{-1} + x(2+k)TZ^{-2} + \ldots$$

therefore

$$Z^{-k} Y(Z) = x(k)TZ^{-k} + x(1+k)TZ^{-(1+k)} + x(2+k)TZ^{-(2+k)} + \ldots$$

that is

$$Z^{-k}Y(Z) = \sum_{n=k}^{\infty} x(n)TZ^{-n}$$

and

$$Z^{-k}Y(Z) + \sum_{n=0}^{k-1} x(n)TZ^{-n} = \sum_{n=k}^{\infty} x(n)TZ^{-n} + \sum_{n=0}^{k-1} x(n)TZ^{-n}$$

Therefore

$$Z^{-k}Y(Z) + \sum_{n=0}^{k-1} x(n)TZ^{-n} = \sum_{n=0}^{\infty} x(n)TZ^{-n} = X(Z)$$

Therefore

$$Y(Z) = Z^k X(Z) - \sum_{n=0}^{k-1} x(n)TZ^{-(n-k)}, \quad \text{for} \quad |Z| > R \qquad (1.12)$$

where R is the radius of convergence of $Z[x(n)T]$.

Example 1.8
Suppose that a signal is defined as $x(n)T = e^{-anT}$; determine the Z-transform of $x(n + 2)T$.

SOLUTION
Firstly we know from studying example 1.2 that for $x(n)T = e^{-anT}$ the corresponding Z-transform is $X(Z) = Z/(Z - e^{-aT})$ for $|Z| > e^{-aT}$. In this example $x(n + 2)T = e^{-a(n+2)T}$, therefore using equation 1.12 we may write

$$Y(Z) = Z^2 \frac{Z}{Z - e^{-aT}} - x(0)TZ^2 - x(1)TZ^1$$

$$= \frac{Z^3}{Z - e^{-aT}} - e^{-0T}Z^2 - e^{-aT}Z$$

Therefore

$$Y(Z) = \frac{(Z e^{-2aT})}{(Z - e^{-aT})} \quad \text{for} \quad |Z| > e^{-aT}$$

(d) Convolution–Summation Property
The input and output signals of a digital filter are related to each other through the convolution–summation property. Now referring to figure 1.8 and using this property we may write

$$y(n)T = g(0)Tx(n)T + g(1)Tx(n - 1)T + g(2)Tx(n - 2)T + \ldots$$

where $g(i)T$ is the *weighting sequence* of the filter.

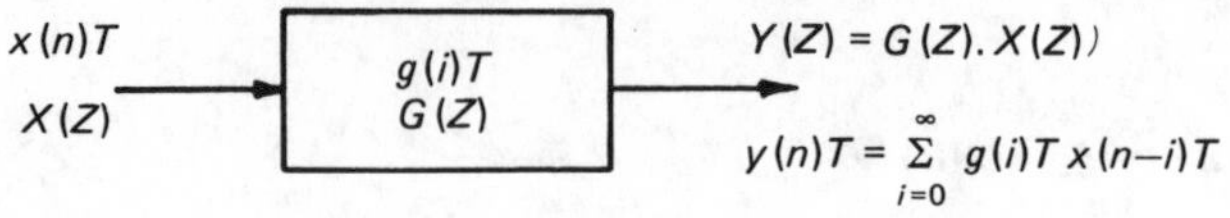

Figure 1.8 Digital filter convolution–summation property

Using equation 1.6 we obtain

$$Y(Z) = \sum_{n=0}^{\infty} [g(0)Tx(n)T + g(1)Tx(n-1)T + g(2)Tx(n-2)T + \ldots]Z^{-n}$$

Therefore

$$Y(Z) = g(0)T \sum_{n=0}^{\infty} x(n)TZ^{-n} + g(1)T \sum_{n=0}^{\infty} x(n-1)TZ^{-n} + \ldots$$

that is

$$Y(Z) = g(0)TX(Z) + g(1)TZ^{-1}X(Z) + g(2)TZ^{-2}X(Z) + \ldots$$
$$= [g(0)T + g(1)TZ^{-1} + g(2)TZ^{-2} + \ldots]\ X(Z)$$

Therefore

$$Y(Z) = G(Z)\ X(Z) \tag{1.13}$$

The ratio $Y(Z)/X(Z)$, equal to $G(Z)$, is commonly referred to as the *pulse transfer function* of the digital filter.

Example 1.9

(a) The pulse transfer function, $G(Z)$, of a digital filter is
$G(Z) = Z/(Z^2 + 2Z - 1)$. Determine the Z-transform of the filter's unit-step response.

(b) Using the convolution–summation property determine the value of the unit-step response at $n = 3$ for the digital filter implemented by

$$y(n)T = 0.1x(n)T + 0.9y(n-1)T$$

SOLUTION

(a) For a unit-step input $X(Z) = Z/(Z-1)$, therefore using equation 1.13 we obtain

$$Y(Z) = \left[\frac{Z}{(Z-1)}\right]\left[\frac{Z}{(Z^2+2Z-1)}\right]$$

that is

$$Y(Z) = \frac{Z^2}{(Z^3 + Z^2 - 3Z + 1)}$$

(b) Using the convolution–summation property the relationship between the input and output signals is

$$y(n)T = \sum_{i=0}^{\infty} g(i)Tx(n - i)T$$

Therefore

$$y(3)T = \sum_{i=0}^{\infty} g(i)Tx(3 - i)T$$

The weighting sequence (impulse response) of the filter is obtained as follows. The linear difference equation of the filter is

$$y(n)T = 0.1x(n)T + 0.9y(n - 1)T$$

Therefore

$$Y(Z)[1 - 0.9Z^{-1}] = 0.1X(Z)$$

Therefore

$$G(Z) = \frac{0.1}{(1 - 0.9Z^{-1})} = 0.1 + 0.09Z^{-1} + 0.081Z^{-2} + \ldots$$

and the coefficients of $G(Z)$ are the impulse response values. Therefore

$$g(i)T = 0.1(0.9)^i \quad \text{for} \quad i = 0, 1, 2, \ldots$$

Therefore

$$y(3)T = 0.1 + 0.1(0.9) + 0.1(0.9)^2 + 0.1(0.9)^3$$

$$= 0.3439$$

(e) Summation Property

Consider a given sequence, $x(n)T$, and suppose that another sequence is generated using the following relationship

$$g(n)T = \sum_{i=0}^{n} x(i)T \quad \text{for} \quad n = 0,1,2,\ldots \tag{1.14}$$

it follows that

$$g(n - 1)T = x(0)T + x(1)T + \ldots + x(n - 1)T$$

But

$$g(n)T = x(0)T + x(1)T + \ldots + x(n - 1)T + x(n)T$$

Therefore

$$g(n)T - g(n-1)T = x(n)T \qquad (1.15)$$

Now using equation 1.6 we obtain

$$X(Z) = \sum_{n=0}^{\infty} [g(n)T - g(n-1)T]Z^{-n}$$

$$= G(Z) - Z^{-1}G(Z)$$

$$= G(Z)\left[\frac{(Z-1)}{Z}\right]$$

Therefore

$$G(Z) = \left[\frac{Z}{(Z-1)}\right] X(Z) \quad \text{for} \quad |Z| > \max(1, R) \qquad (1.16)$$

Example 1.10
Determine the Z-transform of the sequence $g(n)T = n - 1$.

SOLUTION
To satisfy equation 1.15, $x(n)T = 1$ for $n \geqslant 2$, and $x(n)T = 0$ for $n < 2$. That is, $x(n)T$ is the discrete unit-step function delayed by two sampling periods. The Z-transform of a unit-step function is $Z/(Z-1)$, therefore using the right shifting property (equation 1.11) the Z-transform of the delayed unit-step function is $X(Z) = Z^{-2} \times Z/(Z-1) = Z^{-1}/(Z-1)$. Therefore using equation 1.16 we obtain

$$\mathbf{Z}[g(n)T] = G(Z) = Z/(Z-1) \times Z^{-1}/(Z-1)$$

Therefore

$$G(Z) = \frac{1}{(Z-1)^2} \quad \text{for} \quad |Z| > 1$$

(f) Multiplication by A^n Property
Consider a given sequence, $x(n)T$, and suppose that another sequence is generated using the following relationship

$$g(n)T = A^n x(n)T \quad \text{for} \quad n = 0,1,2,\ldots$$

$$\mathbf{Z}[g(n)T] = \mathbf{Z}[A^n x(n)T]$$

$$= \sum_{n=0}^{\infty} A^n x(n)TZ^{-n}$$

$$= \sum_{n=0}^{\infty} x(n)T[A^{-1}Z]^{-n} \quad \text{for} \quad |A^{-1}Z| > R \quad \text{or} \quad |Z| > |A|R$$

Therefore

$$\mathbb{Z}[A^n x(n)T] = X(A^{-1}Z) = X(Z)|_{Z=A^{-1}Z} \tag{1.17}$$

Example 1.11
Determine the Z-transform of $A^n \sin n\omega T$.

SOLUTION
From example 1.3 we know that $\mathbb{Z}[\sin n\omega T]$ is

$$X(Z) = \frac{Z \sin \omega T}{(Z^2 - 2Z \cos \omega T + 1)}$$

therefore using equation 1.17 we obtain

$$\mathbb{Z}[A^n \sin n\omega T] = \frac{A^{-1}Z \sin \omega T}{[(A^{-1}Z)^2 - 2A^{-1}Z \cos \omega T + 1]}$$

$$= \frac{AZ \sin \omega T}{(Z^2 - 2AZ \cos \omega T + A^2)}$$

(g) Periodic Sequence Property
Suppose that a periodic sequence repeats every N discrete-time period, such that

$$x(n)T = x(n - mN)T, \quad \text{for} \quad m = 0,1,2, \ldots$$

Using equation 1.6 we obtain

$$\mathbb{Z}[x(n - mN)T] = \sum_{n=0}^{\infty} x(n - mN)TZ^{-n}$$

$$= \sum_{n=0}^{\infty} x(n)TZ^{-mN}Z^{-n}$$

$$= X(Z) \sum_{m=0}^{\infty} Z^{-mN}$$

but

$$\sum_{m=0}^{\infty} Z^{-mN} = \sum_{m=0}^{\infty} (Z^{-N})^m$$

which in closed form equals $Z^N/(Z^N - 1)$ for $|Z^{-N}| < 1$. Therefore

$$X_p(Z) = \mathbb{Z}[x(n - mN)T] = [Z^N/(Z^N - 1)] \, X(Z) \tag{1.18}$$

Example 1.12
Determine the Z-transform of the periodic sequence shown in figure 1.9.

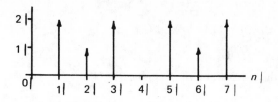

Figure 1.9 Periodic sequence for example 1.12

SOLUTION

The Z-transform of the first period is given by equation 1.6, that is

$$X(Z) = 2Z^{-1} + Z^{-2} + 2Z^{-3}$$

and the Z-transform of the periodic sequence is given by equation 1.18, that is

$$X_p(Z) = [Z^4/(Z^4 - 1)] \; (2Z^{-1} + Z^{-2} + 2Z^{-3})$$
$$= Z(2Z^2 + Z + 2)/(Z^4 - 1)$$

(h) Initial-value Theorem

From equation 1.6 we have

$$X(Z) = x(0)T + x(1)TZ^{-1} + x(2)TZ^{-2} + \ldots$$

by letting $Z^{-1} \to 0$ all terms except $x(0)T$ go to zero. However, $Z^{-1} \to 0$ is the same as $|Z| \to \infty$, therefore

$$x(0)T = \lim_{|Z| \mapsto \infty} X(Z) \tag{1.19}$$

(i) Final-value Theorem

Suppose that we have a finite sequence, $x(n)T$, where $0 \leqslant n \leqslant N$, then using this knowledge we may write

$$\textbf{Z}\,[x(n)T] = X(Z) = \sum_{n=0}^{N} x(n)TZ^{-n}$$

(see equation 1.6). Similarly, for this same sequence delayed by one sampling period we have

$$\textbf{Z}\,[x(n-1)T] = X'(Z) = Z^{-1}X(Z) = \sum_{n=0}^{N-1} x(n)TZ^{-(n+1)}$$

Let us now consider $[X(Z) - X'(Z)] \,|_{Z=1}$, that is

$$\left[\sum_{n=0}^{N} x(n)TZ^{-n} - \sum_{n=0}^{N-1} x(n)TZ^{-(n+1)} \right] \Bigg|_{Z=1} = x(N)T$$

and if $N \to \infty$, then in the limit since N and $N - 1$ converge to the same value it follows that

$$x(\infty)T = \lim_{Z \to 1} [X(Z) - Z^{-1}X(Z)]$$

Therefore

$$x(\infty)T = \lim_{Z \to 1} [(1 - Z^{-1})] X(Z) \qquad (1.20)$$

Example 1.13
Suppose that $X(Z) = 1/(1 - 1.5Z^{-1} + 0.5Z^{-2})$. Determine $x(0)T$ and $x(\infty)T$.

SOLUTION

$$x(0)T = \lim_{Z^{-1} \to 0} X(Z) = 1$$

$$X(Z) = \frac{1}{[(1 - Z^{-1})(1 - 0.5Z^{-1})]}$$

$$x(\infty)T = \lim_{Z \to 1} \left[\frac{(1 - Z^{-1})}{(1 - Z^{-1})(1 - 0.5Z^{-1})} \right]$$

$$= \lim_{Z \to 1} \left[\frac{1}{(1 - 0.5Z^{-1})} \right]$$

$$= \lim_{Z \to 1} \left[\frac{Z}{(Z - 0.5)} \right] = 2$$

Up to this point we have seen how to determine a number of Z-transforms using the infinite series defined by equation 1.6, together with associated properties. An alternative method of determining Z-transforms directly from Laplace transforms is possible, and this method is discussed and demonstrated in the following section.

1.4 METHOD OF DETERMINING Z-TRANSFORMS FROM LAPLACE TRANSFORMS

This method relies on the fact that the Laplace transform of the product of two signals is the *convolution* of their Laplace transforms.

Referring to the impulse modulator discussed in section 1.2, which is defined by equation 1.2, we know that for a physically realisable system

$$\delta_T(t) = \sum_{n=0}^{\infty} \delta(t - nT)$$

Therefore

$$\mathcal{L}\left[\delta_T(t)\right] = \delta_T(p) = \sum_{n=0}^{\infty} e^{-nTp}$$

and therefore in closed form

$$\delta_T(S - p) = \frac{1}{(1 - e^{-(S-p)T})} \tag{1.21}$$

Now denoting the Laplace transform of $x(t)$ as $X(p)$ we may write

$$\mathcal{L}\left[x(t)\,\delta_T(t)\right] = \mathcal{L}\left[x^*(t)\right] = X^*(S)$$

$$= \frac{1}{j2\pi} \int_{c-j}^{c+j} X(p)\,\delta_T(S - p)\,dp \tag{1.22}$$

and substituting equation 1.21 in equation 1.22 we obtain

$$X^*(S) = \frac{1}{j2\pi} \int_{c-j}^{c+j} X(p)\,\frac{1}{(1 - e^{-(S-p)T})}\,dp \tag{1.23}$$

The integral in equation 1.23 can be evaluated using Cauchy's residue theorem yielding

$$X^*(S) = \sum_{\text{poles of } X(p)} \text{residues of} \left[\frac{X(p)}{(1 - e^{-(S-p)T})}\right] \tag{1.24}$$

But for the Z-transform we know that $Z^{-1} = e^{-ST}$ (see section 1.3), therefore equation 1.24 can be rewritten as follows

$$X(Z) = \sum_{\text{poles of } X(p)} \text{residues of} \left[\frac{X(p)}{(1 - e^{pT}Z^{-1})}\right] \tag{1.25}$$

For a pole of order m at $p = x$, the residue used in equation 1.25 is determined by the following equation

$$\text{residue} = \frac{1}{(m-1)!} \lim_{p \to x} \left\{ \frac{d^{m-1}}{dp^{m-1}} \left[(p - x)^m X(p)\,\frac{Z}{Z - e^{pT}} \right] \right\} \tag{1.26}$$

Example 1.14
Using the residue method determine the Z-transform of: (a) $X(p) = k/p(p + 1)$; and (b) $X(p) = 1/p^2$.

SOLUTION
(a) $X(p) = k/p(p + 1)$ has two poles of order $m = 1$ at $p = 0$ and $p = -1$. The corresponding residues are determined using equation 1.26, as follows.

Residue for pole at $p = 0$ is

$$\frac{1}{0!} \lim_{p \to 0} \left\{ \frac{d^0}{dp^0} \left[(p - 0)^1 \frac{k}{p(p + 1)} \frac{Z}{Z - e^{pT}} \right] \right\}$$

$$= \frac{kZ}{Z - 1}$$

Residue for the pole at $p = -1$ is

$$\frac{1}{0!} \lim_{p \to -1} \left\{ \frac{d^0}{dp^0} \left[(p + 1)^1 \frac{k}{p(p + 1)} \frac{Z}{Z - e^{pT}} \right] \right\}$$

$$= -\frac{kZ}{Z - e^{-T}}$$

Therefore

$$X(Z) = kZ \left[\frac{1}{Z - 1} - \frac{1}{Z - e^{-T}} \right]$$

(b) $X(p) = 1/p^2$ has poles of order $m = 2$ at $p = 0$; the corresponding residue is

$$\frac{1}{(2 - 1)!} \lim_{p \to 0} \left\{ \frac{d}{dp} \left[p^2 \frac{1}{p^2} \frac{Z}{Z - e^{pT}} \right] \right\}$$

$$= \lim_{p \to 0} \left\{ \frac{d}{dp} \left[\frac{Z}{Z - e^{pT}} \right] \right\}$$

$$= \lim_{p \to 0} \left[\frac{ZT e^{pT}}{(Z - e^{pT})^2} \right]$$

therefore

$$X(Z) = \frac{ZT}{(Z - 1)^2}$$

1.5 THE INVERSE Z-TRANSFORM

The inversion of the Z-transform is carried out to convert the frequency-domain description of the digital filter to a corresponding time-domain description.

In general, a discrete-time signal or system can be expressed as a ratio of two polynomials in Z, that is, $X(Z) = W(Z)/Y(Z)$. It is possible to divide out $X(Z)$ thereby producing a series expansion, namely

$$X(Z) = a_0 Z^0 + a_1 Z^{-1} + a_2 Z^{-2} + \dots \tag{1.27}$$

where the coefficients a_n are the values of $x(n)T$. Now multiplying both sides of equation 1.27 by Z^{n-1} yields

$$X(Z) Z^{n-1} = a_0 Z^{n-1} + a_1 Z^{n-2} + a_2 Z^{n-3} + \dots \tag{1.28}$$

For equation 1.28 we may take a contour integration round a path enclosing all the poles, that is

$$\oint X(Z) \, Z^{n-1} \, dZ = \oint [a_0 \, Z^{n-1} + a_1 Z^{n-2} + \ldots + a_n Z^{-1} + a_{n-1} Z^0 + \ldots] \, dZ$$

However

$$\oint a_n Z^n \, dZ \begin{cases} = 0 \text{ when } n \neq -1 \\ = j2\pi a_n \text{ when } n = -1 \end{cases}$$

Therefore

$$a_n = x(n)T = \frac{1}{j2\pi} \oint X(Z) \, Z^{n-1} \, dZ \qquad (1.29)$$

The contour integration may be evaluated by the method of residues, thus

$$x(n)T = \sum_{\text{at all poles}} \text{residues of } [X(Z) \, Z^{n-1}] \qquad (1.30)$$

For a pole of order m at $Z = x$, the residue used in equation 1.30 is determined by the following equation

$$\text{residue} = \frac{1}{(m-1)!} \lim_{Z \to x} \left\{ \frac{d^{m-1}}{dZ^{m-1}} \left[(Z - x)^m X(Z) \, Z^{n-1} \right] \right\} \qquad (1.31)$$

Example 1.15

Using the residue method determine the inverse Z-transform of

$$X(Z) = \frac{Z(1 - e^{-T})}{[(Z-1)(Z - e^{-T})]}$$

SOLUTION

$X(Z)$ has two poles of order $m = 1$ at $Z = 1$ and $Z = e^{-T}$. The corresponding residues are determined using equation 1.31, as follows.

Residue for pole at $Z = 1$ is

$$\frac{1}{0!} \lim_{Z \to 1} \left\{ \frac{d^0}{dZ^0} \left[(Z-1)^1 \frac{Z(1 - e^{-T})}{(Z-1)(Z - e^{-T})} Z^{n-1} \right] \right\}$$

$$= \lim_{Z \to 1} \left[\frac{Z(1 - e^{-T})}{(Z - e^{-T})} Z^{n-1} \right] = 1$$

Residue for pole at $Z = e^{-T}$ is

$$\frac{1}{0!} \lim_{Z \to e^{-T}} \left\{ \frac{d^0}{dZ^0} \left[(Z - e^{-T}) \frac{Z(1 - e^{-T})}{(Z-1)(Z - e^{-T})} Z^{n-1} \right] \right.$$

$$= \lim_{Z \to e^{-T}} \left[\frac{Z(1 - e^{-T}) Z^{n-1}}{Z - 1} \right] = -e^{-nT}$$

Now substituting the residues in equation 1.30 we obtain

$$x(n)T = 1 - e^{-nT}$$

At this point let us consider the $X(Z)$ of example 1.15 with a sampling period $T = 1$ s; in this case we have

$$X(Z) = \frac{0.632Z}{[(Z - 1)(Z - 0.368)]}$$

$$= \frac{0.632Z}{(Z^2 - 1.368Z + 0.368)}$$

Now dividing the numerator by the denominator using long division we have

$$
\begin{array}{r}
0.632Z^{-1} + 0.865Z^{-2} + 0.950Z^{-3} + \ldots \\
Z^2 - 1.368Z + 0.368 \overline{\smash{\big)}\, 0.632\,Z} \\
0.632Z - 0.865 + 0.233\,Z^{-1} \\
\overline{ \quad 0.865 - 0.233\,Z^{-1}} \\
0.865 - 1.183\,Z^{-1} + 0.318Z^{-2} \\
\overline{ \quad 0.950Z^{-1} - 0.318Z^{-2}}
\end{array}
$$

therefore

$$X(Z) = 0.632Z^{-1} + 0.865Z^{-2} + 0.952Z^{-3} + \ldots,$$

and the values of $x(n)T$ at the sampling instants are

$$
\left.
\begin{aligned}
x(1)T &= 0.632 \\
x(2)T &= 0.865 \\
x(3)T &= 0.950
\end{aligned}
\right\}
\qquad (1.32)
$$

etc.

From the solution to example 1.15 we see that $x(n)T = 1 - e^{-nT}$, and for $T = 1$ s, $x(n)T = 1 - e^{-n}$. Now taking $n = 2$ we obtain $x(2)T = 1 - e^{-2} = 0.865$, which is identical to the value in equation 1.32. Therefore it is seen that inversion of Z-transforms may be achieved using the long division method demonstrated above. However, it must be understood that this method only yields specific coefficient values and not an expression for the general term of the sequence $x(n)T$.

An alternative method of determining the inverse Z-transform is to expand $X(Z)$ into partial fractions, and then refer directly to a table of Z-transform pairs (see table 1.2) to obtain the corresponding $x(n)T$ expression. The following worked example illustrates the method.

Example 1.16
Using the partial-fraction expansion method determine the inverse Z-transform of

$$X(Z) = \frac{Z(1 - e^{-T})}{[(Z - 1)(Z - e^{-T})]}$$

SOLUTION

$$X(Z) = \frac{Z(1 - e^{-T})}{(Z - 1)(Z - e^{-T})} = \frac{AZ}{Z - 1} + \frac{BZ}{Z - e^{-T}}$$

Therefore

$$Z(1 - e^{-T}) = AZ(Z - e^{-T}) + BZ(Z - 1)$$

equating coefficients of Z^2: $0 = A + B$, and therefore

$$A = -B \tag{1.33}$$

equating coefficients of Z

$$1 - e^{-T} = -(A e^{-T} + B) \tag{1.34}$$

Substituting equation 1.33 in equation 1.34

$$1 - e^{-T} = -B(1 - e^{-T})$$

Therefore $B = -1$ and $A = 1$, and it follows that $X(Z) = Z/(Z - 1) - Z/(Z - e^{-T})$. Now referring to table 1.2 we can directly look up the inverse Z-transform of $X(Z)$, that is

$$x(n)T = 1 - e^{-nT}$$

Compare this with the solution for example 1.15.

1.6 THE DIGITAL FILTER TRANSFER FUNCTION

It has already been established in section 1.3.1(d) that a digital filter may be represented by a linear difference equation which relates the filter's input and output signals. Furthermore, it has also been established that the pulse transfer function of the filter is the ratio $Y(Z)/X(Z)$, where $Y(Z)$ and $X(Z)$ are the Z-transforms of the sampled-data output and input signals respectively (see equation 1.13).

In general, the pulse transfer function, $G(Z)$, is given by

$$G(Z) = \frac{a_0 + a_1 Z^{-1} + a_2 Z^{-2} + \ldots + a_n Z^{-n}}{1 + b_1 Z^{-1} + b_2 Z^{-2} + \ldots + b_m Z^{-m}} = \frac{Y(Z)}{X(Z)} \qquad (1.35)$$

The pulse transfer function representation of equation 1.35 provides us with a useful method for determining the filter's response to various input signals. This can be achieved by first obtaining the Z-transform of the input signal (see section 1.3) yielding $X(Z)$. Next the pulse transfer function, $G(Z)$, is selected or designed (see chapter 2 and chapter 3) and this is multiplied by $X(Z)$ to give the Z-transform of the filter's output response, that is, $Y(Z) = X(Z) G(Z)$. Finally the inverse Z-transform of $Y(Z)$ is obtained (see section 1.5) to yield $y(n)T$.

Now rearranging equation 1.35 we obtain

$$X(Z) [a_0 + a_1 Z^{-1} + a_2 Z^{-2} + \ldots + a_n Z^{-n}] =$$

$$Y(Z) [1 + b_1 Z^{-1} + b_2 Z^{-2} + \ldots + b_m Z^{-m}]$$

and since Z^{-k} implies a time delay equal to k sampling periods [see section 1.3.1(b)] then it follows that the input and output sampled-data signals are related by a linear difference equation as follows

$$\sum_{i=0}^{n} a_i x(h - i)T = \sum_{i=0}^{m} b_i y(h - i)T$$

where $b_0 = 1$ (1.36)

Let us now consider the case when the input signal, $x^*(t)$, is a unit-impulse sequence and equals $\delta(n)T$; we see from inspection of table 1.2 that $X(Z) = 1$, and therefore, referring to equation 1.35, $Y(Z)$ is then seen to be equal to $G(Z)$. Hence we see that the Z-transform of the filter's unit-impulse response is equal to the pulse transfer function. Conversely the inverse Z-transform of the pulse transfer function yields the impulse response (weighting sequence), $g(i)T$, of the filter. This impulse response may then be used in the convolution–summation represent-ation of the filter, as follows

$$y(n)T = \sum_{i=0}^{\infty} g(i)Tx(n - i)T \qquad (1.37)$$

Example 1.17
A digital filter has a pulse transfer function, $G(Z) = Z/(Z - 0.5)$. Determine

(a) a general expression for the filter's unit-step response, and calculate the output values at the first, second and third sampling instants, and
(b) a general expression for the filter's unit-impulse response, using it in the convolution–summation representation method to verify the unit-step response values calculated in part (a).

SOLUTION

(a) For a unit-step input $X(Z) = Z/(Z-1)$

$$Y(Z) = X(Z)G(Z) = \frac{Z^2}{[(Z-1)(Z-0.5)]}$$

and using partial-fraction expansion we obtain

$$Y(Z) = \frac{2Z}{(Z-1)} - \frac{Z}{(Z-0.5)}$$

Therefore referring to table 1.2 we obtain the inverse Z-transform, namely $y(n)T = 2 - 0.5^n$.

The filter output values are therefore

$$y(0)T = 1, y(1)T = 1.5, y(2)T = 1.75$$

(b) Referring to table 1.2 the inverse Z-transform of $G(Z)$ is seen to be $g(i)T = 0.5^i$. The convolution–summation representation is therefore

$$y(n)T = \sum_{i=0}^{\infty} 0.5^i x(n-i)T \text{ (see equation 1.37)}$$

For a unit-step input $x(n-i)T = 1$ for $n \geqslant 0$, and is zero otherwise, therefore

$$y(0)T = 1, \; y(1)T = 1.5, \; y(2)T = 1.75$$

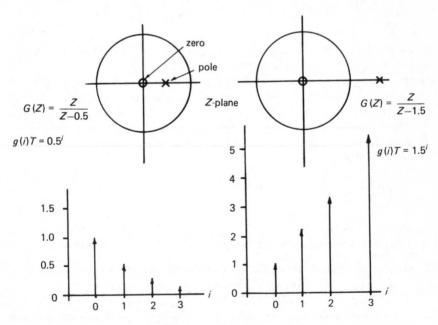

Figure 1.10 Digital filter impulse response governed by location of its poles and zeros

1.6.1 Transfer Function Poles and Zeros

The *poles* and *zeros* of the digital filter may be determined by multiplying the numerator and denominator of $G(Z)$ by Z^m (see equation 1.35), and the resultant numerator and denominator polynomials are then factorised to produce

$$G(Z) = \frac{f(Z - z_1)(Z - z_2)\ldots(Z - z_r)}{(Z - p_1)(Z - p_2)\ldots(Z - p_m)} \tag{1.38}$$

The multiplying factor, f, is a real constant. The elements p_i and z_i are the poles and zeros, respectively, of the digital filter. The poles and zeros are either real or they exist as complex-conjugate pairs.

The behaviour of the digital filter is governed by the location of its poles and zeros in the Z-plane. It is clearly seen from the example shown in figure 1.10 that a different impulse response may be produced by simply moving the pole position. Furthermore the Z-plane pole–zero locations characterise the frequency response of the filter. The pole positions determine whether or not the filter is stable.

1.6.2 Stability Considerations

One form of representation for a digital filter pulse transfer function is to express $G(Z)$ as the ratio of two polynomials in Z. The denominator polynomial, when equated to zero, is the *characteristic equation* of the filter. For a digital filter to be stable all the poles (roots of the characteristic equation) of $G(Z)$ must be within the unit-circle in the Z-plane. If one or more poles are located outside this circle the filter is unstable.

If it is found that one or more poles of $G(Z)$ are located close to the circumference of the unit-circle, thereby rendering $G(Z)$ marginally stable, it will be necessary to investigate how a small change in one of the denominator coefficients, b_k; $1 \leqslant k \leqslant m$ (see equation 1.35), may result in one or more poles moving outside the unit-circle. This small change is brought about by inaccurate representation of the coefficients using finite processor word-length. This later point is discussed more fully in chapter 4.

1.6.3 Steady-state Sinusoidal Response

It is well known that the amplitude and phase response of an electronic system provides valuable information in the design and analysis of transmission circuits, for example, in the design of filter networks. This valuable information can be obtained from the S-plane or Z-plane pole–zero diagram. Referring to equation 1.38 the amplitude response $|G(e^{j\omega T})|$ may be expressed as

$$|G(e^{j\omega T})| = \frac{\displaystyle\prod_{i=1}^{r} \text{vector magnitudes from the } i\text{th zero to the point on the } \omega\text{-axis}}{\displaystyle\prod_{k=1}^{m} \text{vector magnitudes from the } k\text{th pole to the point on the } \omega\text{-axis}}$$

$$\tag{1.39}$$

and the phase response $\angle G(e^{j\omega T})$ may be expressed as

$$\angle G(e^{j\omega T}) = \left\{ \sum_{i=1}^{r} \text{ angles from the } i\text{th zero to the point on the } \omega\text{-axis} \right.$$

$$\left. - \sum_{k=1}^{m} \text{ angles from the } k\text{th pole to the point on the } \omega\text{-axis} \right\} (1.40)$$

The above relationships for $|G(e^{j\omega T})|$ and $\angle G(e^{j\omega T})$ are point-by-point relationships only, that is, vectors must be drawn on the Z-plane from the zeros and poles to every point on the ω-axis, for which the amplitude and phase response is required.

An alternative method of determining the frequency response is to substitute $e^{j\omega T}$ for Z in $G(Z)$, and to compute directly $|G(e^{j\omega T})|$ and $\angle G(e^{j\omega T})$.

Example 1.18
A digital filter has a pulse transfer function $G(Z) = (Z^2 - 0.2Z - 0.08)/(Z^2 - 0.25)$. Determine

(a) the location, in the Z-plane, of the filter's poles and zeros;
(b) whether or not the filter is stable; and
(c) the frequency response of the filter at a frequency equal to one-quarter of the sampling frequency.

SOLUTION
(a) $G(Z) = (Z - 0.4)(Z + 0.2)/(Z - 0.5)(Z + 0.5)$; the values of Z which make $G(Z) = 0$ are $Z = 0.4$ or $Z = -0.2$, therefore the filter has two zeros, one at $Z = 0.4$ and the other at $Z = -0.2$. In contrast, the values of Z which make $G(Z) = \infty$ are $Z = 0.5$ or $Z = -0.5$, therefore the filter has two poles, one at $Z = 0.5$ and the other at $Z = -0.5$ (see figure 1.11).
(b) The two poles of the filter are inside the unit-circle in the Z-plane, therefore the filter is stable.
(c) Substituting $e^{j\omega T}$ for Z in $G(Z)$ we obtain

$$G(e^{j\omega T}) = \frac{e^{j2\omega T} - 0.2 \, e^{j\omega T} - 0.08}{e^{j2\omega T} - 0.25}$$

Therefore

$$G(e^{j\omega T}) = \frac{(\cos 2\omega T - 0.2 \cos \omega T - 0.08) + j(\sin 2\omega T - 0.2 \sin \omega T)}{(\cos 2\omega T - 0.25) + j(\sin 2\omega T)}$$

In this example

$$\omega = \omega_s/4, \text{ but } \omega_s = 2\pi/T, \text{ therefore } \omega T = \pi/2, \text{ and } 2\omega T = \pi.$$

Therefore

$$G(e^{j\omega T}) = \frac{(\cos \pi - 0.2 \cos \pi/2 - 0.08) + j(\sin \pi - 0.2 \sin \pi/2)}{(\cos \pi - 0.25) + j(\sin \pi)}$$

$$= 0.8787 \angle 10.49°$$

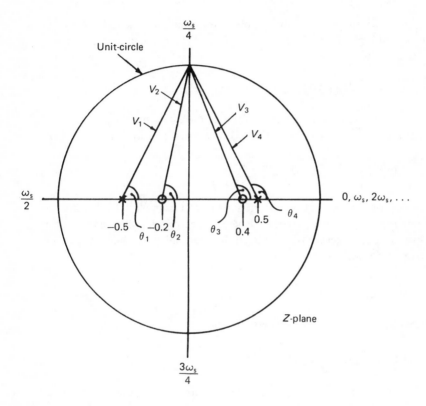

Figure 1.11 Z-plane pole–zero configuration for example 1.18

Now referring to figure 1.11 we have $|V_1| = |V_4| = 1.118$, $|V_2| = 1.020$ and $|V_3| = 1.077$. Also $\theta_1 = 63.435°$, $\theta_2 = 78.690°$, $\theta_3 = 111.801°$ and $\theta_4 = 116.565°$. Therefore using equations 1.39 and 1.40 we obtain

$$|G(e^{j\omega T})| = \frac{|V_2| \cdot |V_3|}{|V_1| \cdot |V_4|} = 0.8787$$

$$\angle G(e^{j\omega T}) = (\theta_2 + \theta_3) - (\theta_1 + \theta_4) = 10.49°$$

Hence it is seen that the values obtained by both methods in part (c) are identical, which therefore demonstrates the validity of both methods.

1.7 THE DISCRETE FOURIER TRANSFORM[9,11]

In general the Z-transform of a sampled-data signal (or system) is given by equation 1.6, that is in the form

$$G(Z) = \sum_{i=0}^{\infty} g(i)TZ^{-i}$$

Let us now suppose that the sampled-data signal (or system) has N distinct values, that is $g(0)T, g(1)T, g(2)T, \ldots, g(N-1)T$, then the corresponding Z-transform is

$$G(Z) = \sum_{i=0}^{N-1} g(i)TZ^{-i}$$

To obtain the frequency spectrum of the sampled-data signal (or system) we substitute $e^{j\omega T}$ for Z (see section 1.63). Therefore we may write

$$G(e^{j\omega T}) = \sum_{i=0}^{N-1} g(i)T\,e^{-ji\omega T}$$

However, we have already seen from our study of example 1.18 that it is impractical to try to evaluate the frequency spectrum for all ω, and therefore only a finite set is generally considered. It is convenient to choose the set of N frequencies defined by

$$\omega_r = \frac{2\pi r}{NT} = \frac{\omega_s r}{N}$$

where $r = 0,1,2, \ldots, (N-1)$. Therefore

$$G(e^{j\omega_r T}) = \sum_{i=0}^{N-1} g(i)T\,e^{-ji2\pi r/N} \tag{1.41}$$

The process of calculating the N values of $G(e^{j\omega_r T})$ is called the *discrete Fourier transform* (DFT).

For notational convenience the definition given by equation 1.41 is often written in the form

$$G_r = \sum_{i=0}^{N-1} g(i)T\,W^{ri} \tag{1.42}$$

$r = 0,1,2, \ldots, (N-1)$, where

$$G_r = G(e^{j\omega_r T}) \quad \text{and} \quad W = e^{-j2\pi/N}$$

However, $\omega_r = \omega_s r/N$ and therefore the DFT is a periodic function in the frequency domain, with a period of $2\pi/T$ rad/s.

It is possible to reverse the transformation to obtain $g(i)T$ values from a sampled frequency spectrum, G_r. This process is achieved by applying the *inverse discrete Fourier transform* (IDFT), which is defined as

$$g(i)T = \frac{1}{N} \sum_{r=0}^{N-1} G_r \, W^{-ri} \qquad (1.43)$$

$i = 0,1,2, \ldots, (N-1)$

Example 1.19
Repeat part (c) of example 1.18 using the DFT method. Comment on the result.

SOLUTION
Firstly it is necessary to determine the impulse response of the filter, $g(i)T$, as follows

$$G(Z) = \frac{(Z - 0.4)\,(Z + 0.2)}{(Z - 0.5)\,(Z + 0.5)}$$

and using partial-fraction expansion we obtain

$$G(Z) = 0.32 + \frac{0.14Z}{Z - 0.5} + \frac{0.54Z}{Z + 0.5}$$

Therefore referring to table 1.2 we obtain the inverse Z-transform. Thus

$$g(i)T = 0.32\delta(i)T + 0.14(0.5)^i + 0.54(-0.5)^i$$

Now substituting $g(i)T$ in equation 1.42 we have

$$G_r = \sum_{i=0}^{N-1} [0.32\delta(i)T + 0.14(0.5)^i + 0.54(-0.5)^i]\,W^{ri}$$

where $r = 0,1, \ldots (N-1)$. We have to evaluate G_r at $\omega_s/4$, therefore we will use $N = 4$, and hence $r = 0,1,2$ and 3. But $\omega_r = \omega_s r/N$, which means that $\omega_1 = \omega_s/4$. Consequently we have to evaluate equation 1.42 for $r = 1$ and $N = 4$, that is

$$G_1 = g(0)TW^0 + g(1)TW^1 + g(2)TW^2 + g(3)TW^3$$

Now $g(0)T = 1, g(1)T = -0.2, g(2)T = 0.17$ and $g(3)T = -0.05$, therefore

$$G_1 = e^{-j0} - 0.2\,e^{-j\pi/2} + 0.17\,e^{-j\pi} - 0.05\,e^{-j3\pi/2}$$

$$= 0.8434\,\angle 10.24°$$

It is seen that this result is slightly in error compared with the actual value of $0.8787\,\angle 10.49°$ and clearly the accuracy may be improved by taking more terms in the DFT. For example, we could let $N = 8$, which means that we would have to evaluate G_2, thus yielding a value of $0.8847\,\angle 10.53°$.

1.8 THE FAST FOURIER TRANSFORM[9,11]

In the previous worked example it was mentioned that accuracy in the DFT calculations could be improved by using 8 terms rather than the 4 terms. In practice it may be desirable to use many more terms, for example, 512 terms or even 1024 terms are not uncommon in digital filtering applications. However, a relatively large number of terms in the DFT does mean that we will probably require the computations to be undertaken by a digital computer.

In using a digital computer to implement the DFT a number of practical advantages can be gained by making $N = 2^y$, where y is a positive integer. In doing this, redundancy in some computations is introduced into the DFT, and by removing the redundant terms computational economy can be improved. This alternative method of obtaining the DFT is called the *fast Fourier transform* (FFT).

To show how the DFT compares with the FFT let us consider the simple case of $y = 3$, that is when $N = 8$. In this case it is convenient to represent r and i (equation 1.42) as binary numbers, that is $r = 4r_2 + 2r_1 + r_0$ and $i = 4i_2 + 2i_1 + i_0$, where r_2, r_1, r_0, i_2, i_1 and i_0 take on values of 0 or 1. The DFT defined by equation 1.42 can be rewritten as

$$G_{(r_2,r_1,r_0)} = \sum_{i_0=0}^{1} \sum_{i_1=0}^{1} \sum_{i_2=0}^{1} g(i_2, i_1, i_0) TW^{(4r_2+2r_1+r_0)(4i_2+2i_1+i_0)}$$

But

$$W^{(4r_2+2r_1+r_0)(4i_2+2i_1+i_0)} =$$

$$W^{(4r_2+2r_1+r_0)4i_2} \times W^{(4r_2+2r_1+r_0)2i_1} \times W^{(4r_2+2r_1+r_0)i_0}$$

Let us now consider each term separately

$$W^{(4r_2+2r_1+r_0)4i_2} = W^{(16r_2+8r_1)i_2} \times W^{(4r_0i_2)} = W^{8(2r_2+r_1)i_2} \times W^{(4r_0i_2)}$$

But $W^8 = (e^{-j2\pi/8})^8$ for $N = 8$, therefore $W^8 = 1$, hence we have

$$W^{(4r_2+2r_1+r_0)4i_2} = W^{4r_0i_2}$$

Similarly

$$W^{(4r_2+2r_1+r_0)2i_1} = W^{8r_2i_1} . W^{(2r_1+r_0)2i_1} = W^{(2r_1+r_0)2i_1}$$

The remaining term $W^{(4r_2+2r_1+r_0)i_0}$ is not reducible.

The DFT may now be written as

$$G_{(r_2,r_1,r_0)} = \sum_{i_0=0}^{1} \sum_{i_1=0}^{1} \sum_{i_2=0}^{1} g(i_2, i_1, i_0) T \, [W^{4r_0i_2} . W^{(2r_1+r_0)2i_1}$$

$$W^{(4r_2+2r_1+r_0)i_0}] \tag{1.44}$$

In equation 1.44 it is convenient to work out each summation separately and to label the intermediate results as follows

$$G_{1(r_0,i_1,i_0)} = \sum_{i_2=0}^{1} g(i_2, i_1, i_0)T\, W^{4r_0 i_2} \tag{1.45}$$

$$G_{2(r_0,r_1,i_0)} = \sum_{i_1=0}^{1} G_{1(r_0,i_1,i_0)}\, W^{(2r_1+r_0)2i_1} \tag{1.46}$$

$$G_{(r_2,r_1,r_0)} = G_{3(r_0,r_1,r_2)} = \sum_{i_0=0}^{1} G_{2(r_0,r_1,i_0)}\, W^{(4r_2+2r_1+r_0)i_0} \tag{1.47}$$

Referring to equation 1.47 we see that the subscript of G is the bit-reversed subscript of G_3, and consequently the final operation in the FFT algorithm is a re-ordering process.

It will now be instructive to take a closer look at equations 1.45, 1.46 and 1.47, as follows. Consider equation 1.45

$$\left. \begin{aligned}
G_{1(0,0,0)} &= g(0,0,0)TW^0 + g(1,0,0)TW^0 \\
G_{1(0,0,1)} &= g(0,0,1)TW^0 + g(1,0,1)TW^0 \\
G_{1(0,1,0)} &= g(0,1,0)TW^0 + g(1,1,0)TW^0 \\
G_{1(0,1,1)} &= g(0,1,1)TW^0 + g(1,1,1)TW^0 \\
G_{1(1,0,0)} &= g(0,0,0)TW^0 + g(1,0,0)TW^4 \\
G_{1(1,0,1)} &= g(0,0,1)TW^0 + g(1,0,1)TW^4 \\
G_{1(1,1,0)} &= g(0,1,0)TW^0 + g(1,1,0)TW^4 \\
G_{1(1,1,1)} &= g(0,1,1)TW^0 + g(1,1,1)TW^4
\end{aligned} \right\} \tag{1.48}$$

However, $W = e^{-j\pi/4}$ for $N = 8$ (see equation 1.42), therefore $W^a = e^{-ja\pi/4}$, and it follows that $W^0 = e^{-j0} = 1$ and $W^4 = e^{-j\pi} = -1$. Therefore equations 1.48 reduce to

$$\left. \begin{aligned}
G_{1(0,0,0)} &= g(0,0,0)T + g(1,0,0)T \\
G_{1(0,0,1)} &= g(0,0,1)T + g(1,0,1)T \\
G_{1(0,1,0)} &= g(0,1,0)T + g(1,1,0)T \\
G_{1(0,1,1)} &= g(0,1,1)T + g(1,1,1)T \\
G_{1(1,0,0)} &= g(0,0,0)T - g(1,0,0)T \\
G_{1(1,0,1)} &= g(0,0,1)T - g(1,0,1)T \\
G_{1(1,1,0)} &= g(0,1,0)T - g(1,1,0)T \\
G_{1(1,1,1)} &= g(0,1,1)T - g(1,1,1)T
\end{aligned} \right\} \tag{1.49}$$

Similarly it can be shown that equations 1.46 reduce to

$$G_{2(0,0,0)} = G_{1(0,0,0)} + G_{1(0,1,0)}$$
$$G_{2(0,0,1)} = G_{1(0,0,1)} + G_{1(0,1,1)}$$
$$G_{2(0,1,0)} = G_{1(0,0,0)} - G_{1(0,1,0)}$$
$$G_{2(0,1,1)} = G_{1(0,0,1)} - G_{1(0,1,1)}$$
$$G_{2(1,0,0)} = G_{1(1,0,0)} + G_{1(1,1,0)}W^2 \qquad (1.50)$$
$$G_{2(1,0,1)} = G_{1(1,0,1)} + G_{1(1,1,1)}W^2$$
$$G_{2(1,1,0)} = G_{1(1,0,0)} - G_{1(1,1,0)}W^2$$
$$G_{2(1,1,1)} = G_{1(1,0,1)} - G_{1(1,1,1)}W^2$$

Similarly it can be shown that equations 1.47 reduce to

$$G_{3(0,0,0)} = G_{2(0,0,0)} + G_{2(0,0,1)}$$
$$G_{3(0,0,1)} = G_{2(0,0,0)} - G_{2(0,0,1)}$$
$$G_{3(0,1,0)} = G_{2(0,1,0)} + G_{2(0,1,1)}W^2$$
$$G_{3(0,1,1)} = G_{2(0,1,0)} - G_{2(0,1,1)}W^2$$
$$G_{3(1,0,0)} = G_{2(1,0,0)} + G_{2(1,0,1)}W^1 \qquad (1.51)$$
$$G_{3(1,0,1)} = G_{2(1,0,0)} - G_{2(1,0,1)}W^1$$
$$G_{3(1,1,0)} = G_{2(1,1,0)} + G_{2(1,1,1)}W^3$$
$$G_{3(1,1,1)} = G_{2(1,1,0)} - G_{2(1,1,1)}W^3$$

By inspecting equations 1.49, 1.50 and 1.51 we see that only 5 separate multiplications are involved, and this should be compared with the number of multiplications involved in the DFT, which for $N = 8$ is 64 (N^2). For the more general case, the N^2 multiplications of the DFT are reduced to $[(N/2)\log_2 N]$ multiplications by using the FFT. For example, if $N = 1024$, the DFT has 1 048 576 complex multiplications, and in contrast the FFT has only 5120, which clearly is a very considerable reduction.

The validity of the FFT method will now be demonstrated by example 1.20.

Example 1.20
Using the FFT repeat part (c) of example 1.18. Comment on the result and the method. Take N to be equal to 8.

SOLUTION
We know from example 1.19 that the impulse response of the filter is

$$g(i)T = 0.32\delta(i)T + 0.14(0.5)^i + 0.54(-0.5)^i$$

Therefore

$g(0)T = 1$

$g(1)T = -0.2$

$g(2)T = 0.17$

$g(3)T = -0.05$

$g(4)T = 0.0425$

$g(5)T = -0.0125$

$g(6)T = 0.010625$

$g(7)T = -0.003125$

For $N = 8$ we have to evaluate G_2 ($\omega_2 = \omega_s/4$), therefore using equations 1.47, 1.49, 1.50 and 1.51 we obtain

$$G_2 = G_{3(2)} = G_{2(2)} + G_{2(3)}W^2$$
$$= [G_{1(0)} - G_{1(2)}] + [G_{1(1)} - G_{1(3)}]\ W^2$$

but $W^2 = -j1$, therefore

$$G_2 = \{[g(0)T + g(4)T] - [g(2)T + g(6)T]\} - j\ \{[g(1)T + g(5)T]$$
$$- [g(3)T + g(7)T]\}$$
$$= 0.861875 + j0.159375 = 0.8765\ \angle 10.48°$$

The answer is close to that quoted at the end of example 1.19. Furthermore, in the above calculations no complex multiplications are involved so the FFT requires far less computer time than the DFT.

The process described above is commonly referred to as the decimation-in-frequency FFT method. In this case the associated basic structure of the FFT is the *butterfly* shown in figure 1.12. Note that the given butterfly is defined such that it has two inputs a and b, which are processed to produce the two outputs x and y, whereby $x = a + (bW^p)$ and $y = a - (bW^p)$.

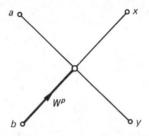

Figure 1.12 Basic FFT butterfly

When considering the complete FFT it is appropriate to simplify the notation used in equations 1.49, 1.50 and 1.51, by replacing the binary number representations with equivalent decimal number representations. For example, rather than writing $g(1,0,1)T$ it is simpler to write $g(5)T$, or more simply $g(5)$. Similarly, we may simplify $G_3(0,1,0)$ to $G_3(2)$. Therefore using this simplified notation, and the fact that in the case considered ($N = 8$) $W^0 = 1$ and $W^4 = -1$, the butterfly processing operations may be summarised by the following sets of equations:

$$
\left.
\begin{aligned}
G_1(0) &= g(0) + g(4) \\
G_1(1) &= g(1) + g(5) \\
G_1(2) &= g(2) + g(6) \\
G_1(3) &= g(3) + g(7) \\
G_1(4) &= g(0) - g(4) \\
G_1(5) &= g(1) - g(5) \\
G_1(6) &= g(2) - g(6) \\
G_1(7) &= g(3) - g(7)
\end{aligned}
\right\}
\qquad (1.52)
$$

$$
\left.
\begin{aligned}
G_2(0) &= G_1(0) + G_1(2) \\
G_2(1) &= G_1(1) + G_1(3) \\
G_2(2) &= G_1(0) - G_1(2) \\
G_2(3) &= G_1(1) - G_1(3) \\
G_2(4) &= G_1(4) + G_1(6)W^2 \\
G_2(5) &= G_1(5) + G_1(7)W^2 \\
G_2(6) &= G_1(4) - G_1(6)W^2 \\
G_2(7) &= G_1(5) - G_1(7)W^2
\end{aligned}
\right\}
\qquad (1.53)
$$

$$\left.\begin{array}{l} G_3(0) = G_2(0) + G_2(1) \\ G_3(1) = G_2(0) - G_2(1) \\ G_3(2) = G_2(2) + G_2(3)W^2 \\ G_3(3) = G_2(2) - G_2(3)W^2 \\ G_3(4) = G_2(4) + G_2(5)W^1 \\ G_3(5) = G_2(4) - G_2(5)W^1 \\ G_3(6) = G_2(6) + G_2(7)W^3 \\ G_3(7) = G_2(6) - G_2(7)W^3 \end{array}\right\} \qquad (1.54)$$

Taking into account the required bit-reversal we may write

$$\left.\begin{array}{l} G(0) = G_3(0) \\ G(1) = G_3(4) \\ G(2) = G_3(2) \\ G(3) = G_3(6) \\ G(4) = G_3(1) \\ G(5) = G_3(5) \\ G(6) = G_3(3) \\ G(7) = G_3(7) \end{array}\right\} \qquad (1.55)$$

The complete eight-point decimation-in-frequency FFT, corresponding to equations 1.52 to 1.55 inclusive, is illustrated by the interconnected butterflies shown in figure 1.13.

An alternative form of FFT structure is the decimation-in-time algorithm. The butterfly shown in figure 1.12 may also be used in this case; however, in implementing the associated FFT computation, the bit-reversed shuffling process is implemented first. For an example, see the complete eight-point decimation-in-time FFT shown in figure 1.14. Note that in this figure the intermediate butterfly computation results are saved in the same program array locations in which the original bit-reversed data was stored. That is, the defined array, $G(\ )$, used to store the bit-reversed input sequence values is used throughout the butterfly computation process, and this same array therefore holds the final output sequence values. This type of process is commonly referred to as an *in-place* computation.

Figure 1.15 shows a flowchart for an N-point in-place decimation-in-time FFT computation, where N and k are positive integers such that

$$N = 2^k, \text{ and } W^p = e^{-j2\pi p/N}$$

A corresponding FORTRAN 77 program is listed in figure 1.16, where the maximum permitted value of N is 1024. Using the eight $g(i)T$ values given in example 1.20, this program yields the results listed in table 1.3. Note that computations for the inverse FFT may be achieved simply by changing the given program so

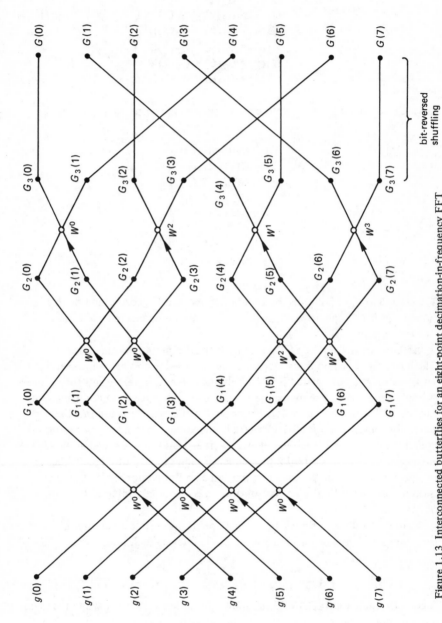

Figure 1.13 Interconnected butterflies for an eight-point decimation-in-frequency FFT

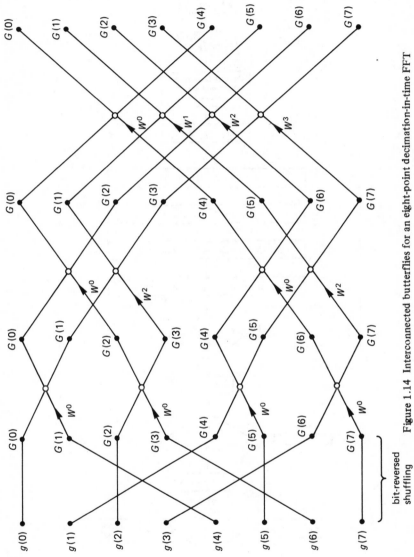

Figure 1.14 Interconnected butterflies for an eight-point decimation-in-time FFT

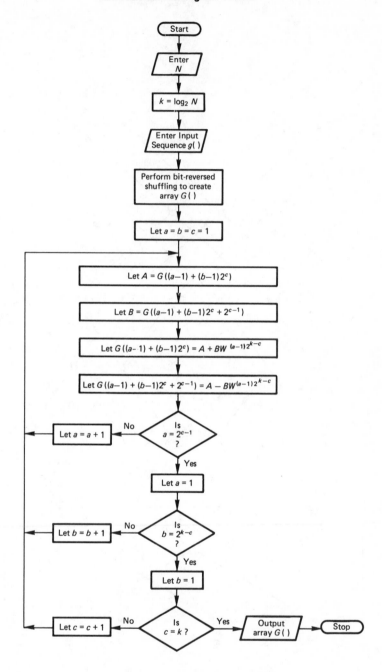

Figure 1.15 Flowchart for an N-point decimation-in-time FFT

```
      PROGRAM DITFFT
C
      COMPLEX G(0:1023),W,A,B,T
      REAL PI
      INTEGER IA,IB,IC,N,I1,I2,K,L,M,I
      INTRINSIC ATAN,ALOG,INT,COS,SIN,CMPLX,REAL,AIMAG
C
C     GET INPUT DATA
C
10    PRINT *,'Enter the number of input data points : N'
      PRINT *,'Maximum value of N is 1024,'
      PRINT *,'N must be equal to 2**K , where K is a positive integer.'
      READ *,N
      IF (N.LE.0 .OR. N.GT.1024) THEN
        PRINT *,'Invalid value of N - try again :'
        GOTO 10
      ENDIF
      K=INT(0.5+ALOG(REAL(N))/ALOG(2.0))
      IF (2**K.NE.N) THEN
        PRINT*,'Invalid value of N - try again :'
        GOTO 10
      ELSE

C
        PRINT *,'Enter the values of the data points one per line e.g.'
        PRINT *,'(real,imaginary)'
        DO 100 I=0,N-1
          READ *,G(I)
100     CONTINUE
C
C     END OF DATA INPUT - NOW REORDER THE INPUT DATA USING BIT REVERSAL
C
        L=1
        DO 300 I=1,N-1
          IF(I.LT.L) THEN
            T=G(L-1)
            G(L-1)=G(I-1)
            G(I-1)=T
          ENDIF
          M=N/2
200       IF (M.LT.L) THEN
            L=L-M
            M=M/2
            GOTO 200
          ENDIF
          L=L+M
300     CONTINUE
C
C     DATA REORDERED - NOW CALCULATE THE TRANSFORM
C
        PI=4.0*ATAN(1.0)
        W=CMPLX(COS(2.*PI/N),-SIN(2.*PI/N))
        DO 600 IC=1,K
          DO 500 IB=1,2**(K-IC)
            DO 400 IA=1,2**(IC-1)
              I1=(IA-1)+(IB-1)*2**IC
              I2=(IA-1)+(IB-1)*2**IC+2**(IC-1)
              A=G(I1)
```

```
              B=G(I2)
              G(I1)=A+B*W**((IA-1)*2**(K-IC))
              G(I2)=A-B*W**((IA-1)*2**(K-IC))
400       CONTINUE
500     CONTINUE
600   CONTINUE
C
C     TRANSFORM COMPLETED - PRINT THE RESULTS AT THE TERMINAL
C
        PRINT '(/,A)','Transform is :-'
        DO 700 I=0,N-1
          WRITE (*,1000)I,REAL(G(I)),AIMAG(G(I))
700     CONTINUE
1000  FORMAT ('G(',I4,') = (',SP,E11.4,',',E11.4,')')
C
        ENDIF
C
C     FINISHED
C
        STOP
        END
```

Figure 1.16   FORTRAN 77 program for the *N*-point decimation-in-time FFT

Table 1.3

```
PROGRAM ENTERED
Enter the number of input data points :
8
Enter the values of the data points one per line e.g.
(real,imaginary)
(1.0,0.0)
(-0.2,0.0)
(0.17,0.0)
(-0.05,0.0)
(0.0425,0.0)
(-0.0125,0.0)
(0.010625,0.0)
(-0.003125,0.0)

Transform is :-
G( 0) = (+0.9575E+00,+0.0000E+00)
G( 1) = (+0.8581E+00,+0.6353E-02)
G( 2) = (+0.8619E+00,+0.1594E+00)
G( 3) = (+0.1057E+01,+0.3251E+00)
G( 4) = (+0.1489E+01,+0.0000E+00)
G( 5) = (+0.1057E+01,-0.3251E+00)
G( 6) = (+0.8619E+00,-0.1594E+00)
G( 7) = (+0.8581E+00,-0.6353E-02)
```

that $-\mathrm{SIN}(2.*\mathrm{PI}/N)$ becomes $+\mathrm{SIN}(2.*\mathrm{PI}/N)$, that is, a change of sign, and by adding the statement $G(I) = G(I)/N$ immediately after the program line: READ *, $G(I)$, or alternatively this additional statement may be inserted immediately after the program line: DO 700 $I = 0, N - 1$.

1.9 TWO-DIMENSIONAL FOURIER TRANSFORM

The computational efficiency of the FFT is even more significant when the signals under consideration are multidimensional. In this context we will restrict our attention specifically to the representation of practical digital images, which are two-dimensional (2-D) signals.

A digital image of extent $N_1 \times N_2$ elements is shown in figure 1.17a, where $g(n_1,n_2)$ represents the magnitude value (grey level value) of the pixel (picture element) at coordinates n_1,n_2. A periodic array of pixels with a horizontal period N_1 and a vertical period N_2 can be expressed as a finite sum of harmonically related complex sinusoids. This is achieved via the inverse 2-D Fourier transform, which is defined as

$$g_p(n_1,n_2) = \frac{1}{N_1 N_2} \sum_{r_1=0}^{N_1-1} \sum_{r_2=0}^{N_2-1} G_p(r_1,r_2)\, W_1^{-n_1 r_1}\, W_2^{-n_2 r_2} \qquad (1.56)$$

where $W_x = e^{-j2\pi/Nx}$, and $G_p(r_1,r_2)$ is the periodic 2-D frequency domain description of the periodic digital image signal, $g_p(n_1,n_2)$, see figure 1.17b.

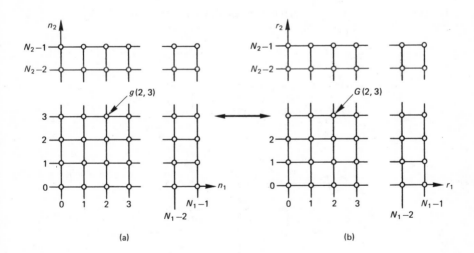

(a)   (b)

Figure 1.17 (a) Digital image pixel array of extent $N_1 \times N_2$. (b) Frequency domain pixel array of extent $N_1 \times N_2$

Because of the invertibility property of the Fourier transform it follows that the 2-D Fourier transform is defined as

$$G_p(r_1, r_2) = \sum_{n_1=0}^{N_1-1} \sum_{n_2=0}^{N_2-1} g_p(n_1, n_2) W_1^{n_1 r_1} W_2^{n_2 r_2}$$

It is advantageous to rewrite the 2-D Fourier transform as

$$G_p(r_1, r_2) = \sum_{n_1=0}^{N_1-1} W_1^{n_1 r_1} \left[\sum_{n_2=0}^{N_2-1} g_p(n_1, n_2) W_2^{n_2 r_2} \right] \qquad (1.57)$$

It may be seen that in equation 1.57 the term inside the square-brackets is a series of N_1 one-dimensional DFTs, each corresponding to a particular value of the n_1 index, which is varied between the limits 0 and $N_1 - 1$, inclusive. We may denote these by

$$G_p(n_1, r_2) = \sum_{n_2=0}^{N_2-1} g_p(n_1, n_2) W_2^{n_2 r_2} \qquad (1.58)$$

Therefore equation 1.58 provides a one-dimensional DFT for each column of the $g_p(n_1, n_2)$ array, see figure 1.18. Now substituting equation 1.58 in equation 1.57 we obtain

$$G_p(r_1, r_2) = \sum_{n_1=0}^{N_1-1} G_p(n_1, r_2) W_1^{n_1 r_1} \qquad (1.59)$$

It is seen that equation 1.59 yields a series of N_2 one-dimensional DFTs, each corresponding to a particular value of the r_2 index, which is varied between the

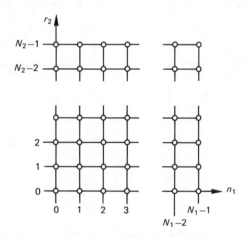

Figure 1.18 One-dimensional column DFTs

limits 0 and $N_2 - 1$, inclusive. Therefore equation 1.59 provides a 1-D DFT for each row of the $G_p(n_1, r_2)$ array, thereby achieving the desired 2-D DFT. Consequently it is seen that the 2-D DFT may be computed as a series of 1-D DFTs, see figure 1.19.

It should be noted that if a 2-D digital image is finite and non-periodic, the pixel array (denoted as $g(n_1, n_2)$) within a period may be considered to be identical to the periodic array $g_p(n_1, n_2)$, but there is a difference in the frequency domain description of each. The Fourier transform of $g_p(n_1, n_2)$ will be a 2-D line spectrum, whereas the Fourier transform of the non-periodic array, $g(n_1, n_2)$, is a continuous 2-D frequency domain description. A practical example of 2-D arrays relating $g(n_1, n_2)$ to $G(r_1, r_2)$ is shown in figure 1.20.

1.10 CONCLUDING REMARKS

This chapter has attempted to provide an introduction to some important basic concepts of sampled-data signals and digital filters. Throughout the chapter it has been assumed that we have no interest in the behaviour of the filter for the time between sampling instants, and this assumption is carried forward in the remainder of the book. It is worth noting that if we want to know what happens to the filter performance between the sampling instants, the well known *modified Z-transform*[6,8,10] should be used to represent the filter's pulse transfer function. However, in an introductory study of digital filters we need not concern ourselves immoderately with the filter's performance between sampling instants, and therefore we do not require any detailed knowledge of the modified Z-transform.

We have seen in this chapter that the standard Z-transform is fundamental to a basic understanding of digital filter concepts; its convolution–summation property gives the relationship between the filter's input and output signals, thereby yielding knowledge of the filter's pulse transfer function, $G(Z)$ (see equation 1.13). Assuming that we have obtained a suitable pulse transfer function, $G(Z)$, it is then possible to represent the filter by its Z-plane pole–zero diagram, or by its linear difference equation. The pole–zero description of the filter yields information concerning the stability of the filter and the steady-state frequency response, and in contrast the linear difference equation describes the time-domain behaviour of the filter. Hence we see that for a given pulse transfer function the filter's characteristic behaviour may be described by either its frequency-domain or time-domain representations.

Naturally a question now arises: how do we derive, or select, a suitable pulse transfer function which will meet a given frequency-domain or time-domain specification? The answer is simply that we must have detailed knowledge of appropriate design methods. The following two chapters deal specifically with digital filter design. In our study of these chapters we will see that the basic concepts demonstrated in this chapter are fundamental to digital filter design and analysis.

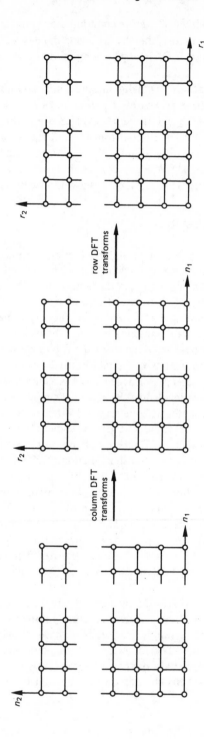

Figure 1.19 Two-dimensional DFT resulting from a series of 1-D column and row DFTs

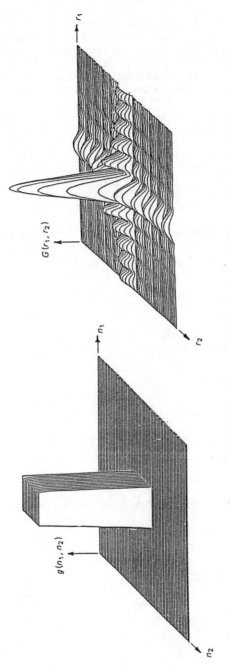

Figure 1.20   Example of a 2–D image signal, $g(n_1, n_2)$, and its Fourier transform, $G(r_1, r_2)$

REFERENCES

1. J. W. Cooley and J. W. Tukey, 'An Algorithm for the Machine Calculation of Complex Fourier Series', *Mathematics of Computing*, 19 (1965) 297–301.
2. J. F. Kaiser, 'Design Methods for Digital Filters', *Proceedings of the First Allerton Conference on Circuit and System Theory*, (1963) 221--36.
3. R. Rabiner and C. M. Rader, *Digital Signal Processing* (IEEE Press, New York, 1972).
4. H. D. Helms and J. F. Kaiser, *Literature in Digital Signal Processing* (IEEE Press, New York, 1975).
5. J. A. Cadzow, *Discrete Time Systems: An Introduction with Interdisciplinary Applications* (Prentice-Hall, Englewood Cliffs, NJ, 1973).
6. J. R. Ragazzini and G. F. Franklin, *Sampled-Data Control Systems* (McGraw-Hill, New York, 1958) chapter 2.
7. P. A. Lynn, *An Introduction to the Analysis and Processing of Signals, Second Edition* (Macmillan, London and Basingstoke, 1982) chapter 4.
8. E. I. Jury, *Theory and Application of the Z-transform Method* (Wiley, New York, 1974).
9. G. D. Bergland, 'A Guided Tour of the Fast Fourier Transform', *IEEE Spectrum*, 6 (1969) 41–52.
10. J. T. Tou, *Digital and Sampled-data Control Systems* (McGraw-Hill, New York, 1959).
11. C. S. Burrus and T. W. Parks, *DFT/FFT Convolution Algorithms: Theory and Implementation* (Wiley, New York and London, 1985).

PROBLEMS

1.1 Write down the Z-transform of the digital filter input sequence $\{2.5, -1.2, -0.08, 8.9, 0.4\}$.

1.2 Determine the Z-transform of

(a) $\dfrac{b - a}{(S + a)(S + b)}$

(b) $\dfrac{\omega}{S^2 + \omega^2}$

1.3 Determine the Z-transform and region of convergence for

$$f(n) = \begin{cases} (\tfrac{1}{4})^n & \text{for} \quad n \geqslant 0 \\ 0 & \text{for} \quad n < 0 \end{cases}$$

1.4 Determine the inverse Z-transform of

(a) $\dfrac{(Z + 3)}{(Z - \frac{1}{4})}$

(b) $\dfrac{Z^2}{[(Z - 0.5)(Z - 1)]}$

1.5 The pulse transfer function of a digital filter is $G(Z) = (Z + 0.5)/(Z + 0.25)$. Determine: (a) a general expression for the filter's unit-step response, and evaluate it at the first four sampling instants; and (b) a general expression for the filter's unit-impulse response, and use it in the convolution–summation representation to verify the unit-step values calculated in part (a).

1.6 A digital filter has a pulse transfer function

$$G(Z) = \frac{Z^2 - 0.05Z - 0.05}{Z^2 + 0.1Z - 0.2}$$

Determine: (a) the location, in the Z-plane, of the filter's poles and zeros; (b) whether or not the filter is stable; (c) a general expression for the filter's unit-impulse response; (d) the filter's linear difference equation; (e) the frequency response of the filter at a frequency equal to one half of the sampling frequency; and (f) the frequency response of the filter obtained via the DFT for $N = 4$.

1.7 Suppose that a continuous signal is sampled to produce sixteen sampled-data values; estimate the percentage reduction in computation time when evaluating the signal's frequency spectrum via the FFT instead of the DFT.

2 Design of Recursive Digital Filters

2.1 INTRODUCTION

Recursive digital filters are commonly referred to as infinite impulse response (IIR) filters. The term *recursive* intrinsically means that the output of the digital filter, $y(n)T$, is computed using the present input, $x(n)T$, and previous inputs and outputs, namely, $x(n-1)T, x(n-2)T, \ldots, y(n-1)T, y(n-2)T, \ldots$, respectively.

The design of a recursive digital filter centres around finding the pulse transfer function, $G(Z)$, which satisfies a given performance specification. This design process involves finding the filter coefficients—the a_is and b_is of $G(Z)$, thereby yielding a pulse transfer function which is a rational function in Z^{-1}.

A number of useful design methods are discussed in this chapter; each one is basically a mathematical method of obtaining a solution to the problem of approximating to a desired filter characteristic. For example, it may be required that the frequency response of the filter approximates a lowpass *brickwall* characteristic, which would involve one of the frequency-domain design methods: bilinear Z-transform, matched Z-transform, frequency sampling, and direct approach using squared magnitude functions. In contrast the design of the recursive digital filter may be viewed in terms of a time-domain specification. For example, in the case of a wave-shaping digital filter it is required to find the weights (impulse response), $g(0)T, g(1)T, \ldots$, which when convolved with the input samples, $x(0)T, x(1)T, \ldots$, produce an output waveform having the desired shape.

The frequency-domain approach to the design of recursive digital filters may be subdivided into two main techniques of solution. The first method is an indirect approach, which requires that a suitable prototype continuous (analogue) filter transfer function, $G(S)$, is designed, and subsequently this is transformed via an appropriate S-plane to Z-plane mapping to give a corresponding digital filter pulse transfer function, $G(Z)$. The mappings used in this chapter are the standard Z-transform (impulse-invariant design method), the bilinear Z-transform and the matched Z-transform. The second method is a direct approach which is concerned with the Z-plane representation of the digital filter, and the derivation of $G(Z)$ is achieved working directly in the Z-plane. This direct approach is used in the design of frequency sampling filters and filters based on squared magnitude functions.

Recursive digital filters are generally more economical in execution time and storage requirements compared with their non-recursive counterparts. However, some types of recursive digital filter have non-linear phase characteristics which may produce unacceptable waveform distortion. The linear phase characteristic of the frequency sampling filters discussed in this chapter, and the fact that this type of filter has integer coefficients makes this an attractive, economical and useful recursive digital filter in some simple applications.

2.2 INDIRECT APPROACH USING PROTOTYPE CONTINUOUS FILTER

In general the design of a prototype continuous (analogue) filter involves two main steps, namely

(1) firstly deriving a realisable transfer function, $G(S)$; and
(2) subsequently synthesising the transfer function.

In contrast, the *indirect approach* to the design of digital filters is concerned, in part, with step (1) only. That is, in using this approach we will firstly obtain a suitable transfer function, $G(S)$, and then derive the corresponding pulse transfer function, $G(Z)$, using one of the Z-transform methods discussed later in this chapter. Thus for the *indirect approach* we have the basic design method illustrated in figure 2.1.

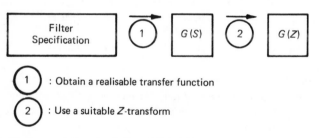

Figure 2.1 Indirect approach to the design of digital filters

In order to carry out step (1), that is, in going from the filter specification to the transfer function $G(S)$, it will be necessary to consider the design of continuous filters, and in particular Butterworth and Chebyshev filters will be briefly reviewed since both types are commonly used in this approach to digital filter design.

2.2.1 Review of Butterworth and Chebyshev Filters[1]

(a) Butterworth Lowpass Filter

For a Butterworth filter to approximate to the ideal lowpass characteristic (see figure 2.2) the following relationship is used

$$|G(j\omega)|^2 = \frac{1}{1 + (\omega/\omega_c)^{2n}} = \frac{1}{1 + (-1)^n S^{2n}}\bigg|_{S=j\omega/\omega_c} \tag{2.1}$$

where $|G(j\omega)|^2$ is the squared magnitude of the filter's transfer function, n is the order of the filter complexity, S is the complex frequency variable (equal to $\sigma + j\omega$: a complex number) and ω_c is the cutoff frequency.

The first $(2n - 1)$ derivatives of an nth-order Butterworth lowpass filter are zero at $\omega = 0$. This property gives the filter a *maximally flat magnitude* characteristic, and as $n \to \infty$ the Butterworth magnitude function approaches the ideal lowpass characteristic shown in figure 2.2.

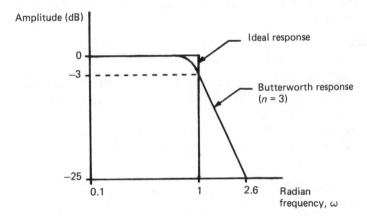

Figure 2.2 Comparison of Butterworth response ($n = 3$) with ideal response

In equation 2.1 the roots of the denominator term, $1 + (-1)^n S^{2n}$, correspond to the poles of the filter which lie equally spaced on the circumference of a unit-circle in the S-plane. Letting the angle of the kth root be denoted by ϕ_k, where $k = 0, 1, 2, \ldots, (2n - 1)$, then

$$\phi_k = k\pi/n \text{ for } n \text{ odd} \tag{2.2}$$

or

$$\phi_k = \frac{(k + \frac{1}{2})\pi}{n} \text{ for } n \text{ even} \tag{2.3}$$

For a stable Butterworth filter the transfer function, $G(S)$ is a rational function having a numerator equal to unity and denominator determined by selecting the roots of $1 + (-1)^n S^{2n}$ that are located in the left-hand half of the S-plane. Note that mirror images of the poles of $G(S)$ exist in the right-hand half of the S-plane, and we associate these with $G(-S)$. To illustrate the method of determining $G(S)$ an example will now be considered.

Example 2.1
Suppose that to meet a given filter specification the order of the filter must be $n = 3$. Derive the corresponding transfer function for a Butterworth lowpass continuous filter.

SOLUTION
Since $k = 0, 1, 2, \ldots, (2n - 1)$, we see that there are six poles on the circumference of the unit-circle in the S-plane, their corresponding angular positions being determined by equation 2.2, and their corresponding S-plane representation

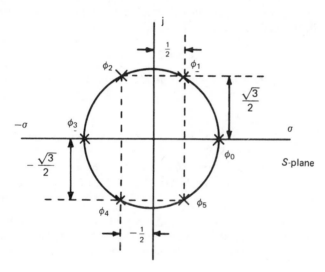

Figure 2.3 S-plane pole positions for third-order Butterworth lowpass filter

is shown in figure 2.3. The poles ϕ_2, ϕ_3 and ϕ_4 are in the left-hand half of the S-plane, and these are the poles used in determining $G(S)$, as follows

$$\phi_2 = -\tfrac{1}{2} + j\ \frac{\sqrt{3}}{2}$$

$$\phi_3 = -1 + j\,0$$

$$\phi_4 = -\tfrac{1}{2} - j\ \frac{\sqrt{3}}{2}$$

Hence

$$G(S) = \cfrac{1}{(S+1)(S+\tfrac{1}{2}+j\tfrac{\sqrt{3}}{2})(S+\tfrac{1}{2}-j\tfrac{\sqrt{3}}{2})}$$

Therefore

$$G(S) = \frac{1}{(S+1)(S^2+S+1)} \qquad (2.4)$$

At this point it will be instructive to recall the basic definition of a pole and to check the expression for $G(S)$, equation 2.4, using it. Hence recalling that the poles of a transfer function are the real or complex values of S which make $G(S)$ infinite, then substitution of ϕ_2 or ϕ_3 or ϕ_4 in equation 2.4 does indeed make $G(S)$ infinite, and therefore for $n=3$ a Butterworth lowpass filter has the transfer function derived in example 2.1, namely equation 2.4. Table 2.1 lists Butterworth polynomials in factored form for $n=1$ to $n=6$.

Table 2.1

n	Butterworth Polynomials (in Factored Form)
1	$(S+1)$
2	$(S^2+\sqrt{2}S+1)$
3	$(S^2+S+1)(S+1)$
4	$(S^2+0.7653S+1)(S^2+1.84776S+1)$
5	$(S+1)(S^2+0.6180S+1)(S^2+1.6180S+1)$
6	$(S^2+0.5176S+1)(S^2+\sqrt{2}S+1)(S^2+1.9318S+1)$

In example 2.1 it was assumed that the value of n was known; however, in practice the order of filter complexity, n, would have to be determined using the data of the filter's specification. With this latter point in mind let us now consider the following

$$\text{attenuation, } -X \text{ dB} = 10\log_{10}|G(j\omega)|^2 \qquad (2.5)$$

Now substituting equation 2.1 in equation 2.5 yields

$$-X \text{ dB} = 10\log_{10}\left[\frac{1}{1+(\omega/\omega_c)^{2n}}\right]$$

$$= 10\log_{10} 1 - 10\log_{10}\left[1+\left(\frac{\omega}{\omega_c}\right)^{2n}\right]$$

Therefore

$$\text{attenuation, } X \text{ dB} = 10\log_{10}\left[1+\left(\frac{\omega}{\omega_c}\right)^{2n}\right] \qquad (2.6)$$

Thus we see that equation 2.6 will yield the value of n provided that the values of X, ω, and ω_c are given in the filter's specification. The method is illustrated in example 2.2 below.

Example 2.2
The specification for a Butterworth lowpass continuous filter reads as follows

(a) cutoff frequency, $\omega_c = 0.75$, and
(b) amplitude response to be at least 20 dB down when $\omega = 3$.

Determine the value of n that will satisfy the given specification.

SOLUTION
Using equation 2.6 we obtain

$$20 = 10 \log_{10} \left[1 + \left(\frac{3}{0.75} \right)^{2n} \right]$$

Therefore

$$\text{antilog}_{10}\, 2 = 1 + (4)^{2n}$$

Therefore

$$99 = (4)^{2n}$$

Therefore

$$\log_{10} 99 = 2n \log_{10} 4$$

Therefore

$$n \approx 1.657$$

However, n must be an integer, therefore increasing $n = 2$ will satisfy the filter's specification since the attenuation will exceed the required 20 dB at $\omega = 3$. In contrast, if we use $n = 1$ the attenuation will be less than 20 dB at $\omega = 3$. Therefore it is advisable to take the next higher integer value for n so that the filter's specification will be satisfied; thus in this example $n = 2$.

From studying example 2.1 and example 2.2 we see that in designing a Butterworth lowpass continuous filter the design process denoted by the symbol: ①, figure 2.1, may be subdivided as illustrated in figure 2.4. Thus figure 2.4 summarises the design process for deriving the transfer function, $G(S)$, for a Butterworth lowpass continuous filter.

(b) Chebyshev Lowpass Filter
Before considering the Chebyshev lowpass filter in detail it is worth recalling that in using the Butterworth approximation the maximally flat characteristic is best at

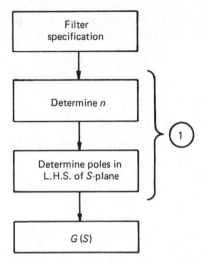

Figure 2.4 Design process for deriving the transfer function of a Butterworth lowpass filter

$\omega = 0$ and as the cutoff frequency is approached the error in the approximation increases, see figure 2.2. Alternatively an approximation having a small (selectable) value of 'ripple' near unity will be equally good at $\omega = 0$ and $\omega = 1$, consequently this type of function is usually referred to as an 'equal ripple approximation'. This type of equal ripple function results from use of Chebyshev cosine polynomials, namely

$$C_n(\omega) = \cos(n \cos^{-1} \omega)|_{|\omega| \leqslant 1}$$

and

$$C_n(\omega) = \cosh(n \cosh^{-1} \omega)|_{|\omega| > 1}$$

Now when $n = 0$ we have $C_0(\omega) = 1$, and for $n = 1$ we have $C_1(\omega) = \omega$. Higher order polynomials may be determined using the recursive formula

$$C_n(\omega) = 2\omega C_{n-1}(\omega) - C_{n-2}(\omega) \tag{2.7}$$

Table 2.2

n	Chebyshev Polynomials $[C_n(\omega)]$
0	1
1	ω
2	$2\omega^2 - 1$
3	$4\omega^3 - 3\omega$
4	$8\omega^4 - 8\omega^2 + 1$
5	$16\omega^5 - 20\omega^3 + 5\omega$
6	$32\omega^6 - 48\omega^4 + 18\omega^2 - 1$

For example, for $n = 2$ we have

$$C_2(\omega) = 2\omega\omega - 1 = 2\omega^2 - 1$$

(see table 2.2).

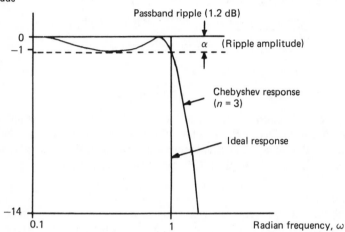

Figure 2.5 Chebyshev lowpass filter response for example 2.3

For a Chebyshev filter to approximate to the ideal lowpass characteristic (see figure 2.5) the following relationship is used

$$|G(j\omega)|^2 = \frac{1}{1 + \epsilon^2 [C_n(\omega)]^2}$$

where ϵ is real and $\ll 1$. Therefore

$$|G(j\omega)| = \frac{1}{\sqrt{1 + \epsilon^2 [C_n(\omega)]^2}}$$

However, in the stopband as ω increases a frequency is reached ($= \omega_{sb}$) where $\epsilon^2 [C_n(\omega)]^2 \gg 1$, so that we may deduce that

$$|G(j\omega)| \approx \frac{1}{\epsilon\, C_n(\omega)} \bigg|_{\omega \geqslant \omega_{sb}}$$

We know the attenuation, $-X$ dB $= 10 \log_{10} |G(j\omega)|^2$, see equation 2.5, therefore attenuation is

$$X \text{ dB} = -20 \log_{10} \frac{1}{\epsilon\, C_n(\omega)}$$

$$= -20 \log_{10} [\epsilon \, C_n(\omega)]^{-1}$$

$$= 20 \log_{10} \epsilon + 20 \log_{10} C_n(\omega)$$

For large values of ω (in stopband), $C_n(\omega)$ can be approximated by the leading term of Chebyshev polynomials (see table 2.2). Therefore attenuation is

$$X \text{ dB} = 20 \log_{10} \epsilon + 20 \log_{10} (2^{n-1} \, \omega^n)$$

$$= 20 \log_{10} \epsilon + 20 \log_{10} 2^{n-1} + 20 \log_{10} \omega^n$$

$$= 20 \log_{10} \epsilon + (n-1) \, 20 \log_{10} 2 + 20n \log_{10} \omega$$

Therefore attenuation is

$$X \text{ dB} \approx 20 \log_{10} \epsilon + 6(n-1) + 20n \log_{10} \omega \qquad (2.8)$$

Clearly the Chebyshev approximation depends on the values of ϵ and n. The maximum permissible ripple fixes the value of ϵ, and once this value of ϵ has been determined the value of the attenuation in the stopband fixes the value of the filter complexity, n. Example 2.3 illustrates the design process.

Example 2.3
A Chebyshev lowpass characteristic is required to have a maximum passband ripple of 1.2 dB and an attenuation of at least 25 dB at $\omega = 2.5$. Determine the values of ϵ and n.

SOLUTION
At $\omega = 1$ the ripple is 1.2 dB and $[C_n(1)]^2 = 1$; using equation 2.5 we have

$$1.2 \text{ dB} = -10 \log_{10} \left[\frac{1}{1 + \epsilon^2} \right]$$

Therefore

$$1.2 \text{ dB} = 10 \log_{10} (1 + \epsilon^2)$$

Therefore

$$\left\{ \left[\text{antilog}_{10} \left(\frac{1.2}{10} \right) \right] - 1 \right\}^{1/2} = \epsilon = 0.5641$$

At $\omega = 2.5$, namely in the stopband, attenuation $= 25$ dB; hence using equation 2.8 we obtain

$$25 = 20 \log_{10} (0.5641) + 6(n-1) + 20n \log_{10} 2.5$$

and solving for n yields $n = 2.577$. However, recalling that n must be an integer, then taking the next higher integer value, $n = 3$, would satisfy the filter's specification.

The ripple amplitude, α, is given by $\alpha = 1 - (1 + \epsilon^2)^{-1/2}$, see figure 2.5, and for example 2.3, namely when $\epsilon = 0.5641$, the corresponding value of α is 0.129.

Having obtained ϵ and n we could then continue to derive $G(S)$; however, this is an involved process and consequently only relevant results need be quoted herein. Firstly a design parameter is defined as follows

$$A_k = \frac{1}{n} \, \sinh^{-1} \left(\frac{1}{\epsilon} \right) \qquad (2.9)$$

It can be shown that a comparison of the normalised Chebyshev pole locations with the normalised Butterworth pole locations reveals that the imaginary parts are identical, and the real part of the Butterworth pole times a factor $\tanh A_k$ is equal to the real part of the Chebyshev pole,[1] see figure 2.6. Hence knowing the normalised Butterworth poles the corresponding normalised Chebyshev poles can be derived. The denormalised Chebyshev poles are obtained by multiplying the normalised Chebyshev poles by a denormalising-factor equal to $\cosh A_k$.

Example 2.4
Derive the transfer function, $G(S)$, corresponding to the specification given in example 2.3.

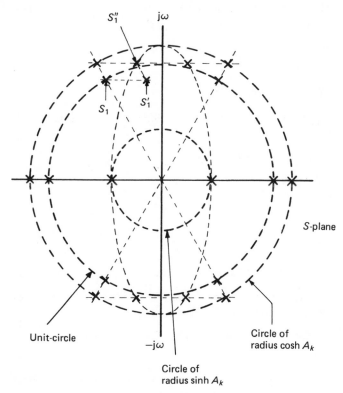

Figure 2.6 Comparison of Butterworth and Chebyshev pole positions

SOLUTION

$$A_k = \tfrac{1}{3} \sinh^{-1}\left(\frac{1}{0.5641}\right) = 0.4457$$

$\tanh A_k = 0.4184$

$\cosh A_k = 1.1009$

For $n = 3$, the Butterworth poles are $\phi_k = \pi k/3$, and $k = 0, 1, 2, 3, 4$ and 5, hence it follows that $\phi_0 = 0$, $\phi_1 = \pi/3$, $\phi_2 = 2\pi/3$, $\phi_3 = \pi$, $\phi_4 = 4\pi/3$ and $\phi_5 = 5\pi/3$. Selecting only the poles in the left-hand half of the S-plane we obtain

$$S_1 = \cos\frac{2\pi}{3} + j\sin\frac{2\pi}{3} = -0.5 + j0.866$$

$$S_2 = \cos\pi + j\sin\pi = -1 + j0$$

$$S_3 = \cos\frac{4\pi}{3} + j\sin\frac{4\pi}{3} = -0.5 - j0.866$$

Multiplying the real parts of S_1, S_2 and S_3 by $\tanh A_k$ we obtain the normalised Chebyshev poles

$$S_1' = -0.2092 + j0.866$$

$$S_2' = -0.4184 + j0$$

$$S_3' = -0.2092 - j0.866$$

Multiplying S_1', S_2' and S_3' by $\cosh A_k$ we obtain the denormalised Chebyshev poles

$$S_1'' = -0.2303 + j0.9534$$

$$S_2'' = -0.4606 + j0$$

$$S_3'' = -0.2303 - j0.9534$$

Therefore

$$G(S) = \frac{f}{(S + 0.4606)(S + 0.2303 - j0.9534)(S + 0.2303 + j0.9534)}$$

the factor f is a multiplying constant which ensures the correct gain at $\omega = 0$, that is, at $\omega = 0$, $|G(j0)| = 1$, therefore

$$f = 1\,[0.4606(0.2303 - j0.9534)(0.2303 + j0.9534)] = 0.4431$$

Therefore

$$G(S) = \frac{0.4431}{(S + 0.4606)(S + 0.2303 - j0.9534)(S + 2303 + j0.9534)} \tag{2.10}$$

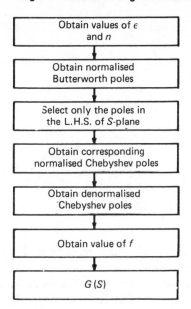

Figure 2.7 Design process for deriving the transfer function of a Chebyshev lowpass filter

Figure 2.7 summarises the recommended design process for deriving $G(S)$ for a Chebyshev lowpass continuous filter.

2.2.2 Impulse-invariant Design Method[2,3,4]

For a continuous filter the impulse response, $g(t)$, is defined as $\mathcal{L}^{-1}[G(S)]$. Similarly, for a digital filter the impulse response, $g(n)T$, is defined as $\mathbf{Z}^{-1}[G(Z)]$; n being an integer corresponding to the nth sampling instant, that is $n = 0,1,2,\ldots$.

Now consider a continuous filter transfer function $G(S)$ with m simple poles, that is

$$G(S) = \sum_{i=1}^{m} \frac{f_i}{(S + S_i)} \tag{2.11}$$

The corresponding impulse response is thus

$$g(t) = \mathcal{L}^{-1}\left[\sum_{i=1}^{m} \frac{f_i}{(S + S_i)}\right] = \sum_{i=1}^{m} f_i\, e^{-S_i t} \tag{2.12}$$

At each sampling instant we want $g(n)T$ to equal $g(t)$, that is $g(n)T = g(t)$ for $t = 0, T, 2T, \ldots$, thus

$$g(n)T = \sum_{i=1}^{m} f_i\, e^{-S_i nT} \tag{2.13}$$

Taking the Z-transform of both sides of equation 2.13 yields

$$G(Z) = \sum_{i=1}^{m} f_i \left[\frac{1}{1 - e^{-S_i T} Z^{-1}} \right] \tag{2.14}$$

Hence we see that

$$G(S) = \sum_{i=1}^{m} \frac{f_i}{(S + S_i)}$$

transforms to

$$\sum_{i=1}^{m} \left[\frac{f_i}{1 - e^{-S_i T} Z^{-1}} \right] = G(Z) \tag{2.15}$$

Thus we see that equation 2.15 is a direct application of the standard Z-transform, which was discussed in chapter 1.

Now at this point it is appropriate to recall that a digital filter input signal will be a set of sampled-data values, and therefore the input signal needs multiplying by a factor equal to the sampling period, T, so that the spectra of $x(n)T$ will adequately represent $x(t)$. This objective is most conveniently accomplished by defining the digital filter transfer function as

$$G(Z) \times T = T \sum_{i=1}^{m} \left[\frac{f_i}{1 - e^{-S_i T} Z^{-1}} \right] \tag{2.16}$$

Hence we see that $G(Z)$ may be obtained by firstly expressing $G(S)$ as a sum of partial fractions and then applying the transform given in equation 2.15. However, the form of equation 2.15 is not easy to handle when S_i is a complex number because correspondingly f_i will also be a complex number, which means further algebraic manipulations will be necessary to reduce $G(Z)$ to a real function of Z.

The following example will be used to illustrate

(1) the methods used in obtaining the amplitude/frequency and phase/frequency characteristics of continuous and digital filters;
(2) the impulse-invariant design method;
(3) the significance of aliasing errors; and
(4) the impulse-invariant property, that is, the preservation of the continuous filter's impulse response at the sampling instants.

Example 2.5
The transfer function of a third-order Chebyshev lowpass filter is the $G(S)$ corresponding to equation 2.10, that is

$$G(S) = \frac{0.4431}{(S + 0.4606)(S + 0.2303 - j0.9534)(S + 0.2303 + j0.9534)}$$

Determine

(a) the amplitude/frequency and phase/frequency characteristics of $G(S)$;

(b) the corresponding impulse-invariant digital filter;

(c) the amplitude/frequency and phase/frequency characteristics of the impulse-invariant digital filter;

(d) the impulse reponse, $g(t)$ of $G(S)$; and

(e) the unit-sample response, $g(n)T$, of the impulse-invariant digital filter (take $T = 1$ s).

Comment on the results.

SOLUTION

(a) The frequency response is obtained by substituting $j\omega$ for S in $G(S)$, thus

$$G(j\omega) = \frac{0.4431}{(0.4431 - 0.9212\omega^2) + j(1.1742\omega - \omega^3)} \qquad (2.17)$$

Using equation 2.17 the amplitude/frequency and phase/frequency characteristics of $G(S)$ may be determined by calculating $|G(j\omega)|$ and $\angle G(j\omega)$

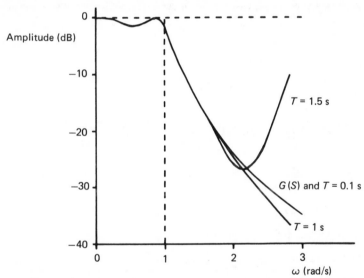

Figure 2.8 Amplitude/frequency response for third-order Chebyshev lowpass filter (example 2.5)

respectively over a suitable normalised-frequency range, for example, $0 \leqslant \omega \leqslant 3$. These characteristics are shown in figure 2.8 and figure 2.9.

(b) In general, the poles of a third-order transfer function will have values

$S_1 = -a$

$S_2 = -b + jc$

$S_3 = -b - jc$

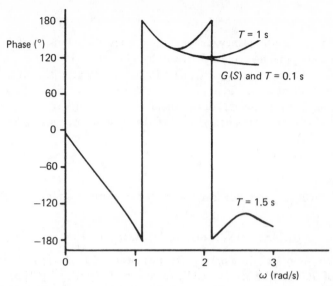

Figure 2.9   Phase/frequency response for third-order Chebyshev lowpass filter (example 2.5)

Therefore

$$G(S) = \frac{f}{(S + a)(S + b - jc)(S + b + jc)}$$

where

$$f = 1 [a(b - jc)(b + jc)] = a(b^2 + c^2)$$

Therefore

$$G(S) = \frac{a(b^2 + c^2)}{(S + a)(S^2 + 2bS + b^2 + c^2)}$$

$$= \frac{a(b^2 + c^2)}{(S + a)[(S + b)^2 + c^2]}$$

and expanding into partial fractions we obtain

$$G(S) = \frac{a(b^2 + c^2)}{(b - a)^2 + c^2} \left[\frac{1}{S + a} - \frac{S + (2b - a)}{(S + b)^2 + c^2} \right]$$

Therefore

$$G(S) = \frac{a(b^2 + c^2)}{(b - a)^2 + c^2} \left\{ \frac{1}{S + a} - \frac{S + b}{(S + b)^2 + c^2} \right.$$

$$\left. - \frac{(b - a)}{c} \left[\frac{c}{(S + b)^2 + c^2} \right] \right\} \qquad (2.18)$$

Now substituting in equation 2.18 the pole values used in equation 2.10 we obtain

$$G(S) = 0.4606 \left\{ \frac{1}{S + 0.4606} - \frac{S + 0.2303}{(S + 0.2303)^2 + 0.9090} + 0.2416 \right.$$

$$\left. \left[\frac{0.9534}{(S + 0.2302)^2 + 0.9090} \right] \right\} \qquad (2.19)$$

Equation 2.19 may be transformed to an impulse-invariant digital filter using the following standard Z-transforms

$$\frac{1}{S + 0.4606} \quad \text{transforms to} \quad \frac{Z}{Z - e^{-0.4606T}} \qquad (2.20)$$

$$\frac{S + 0.2303}{(S + 0.2303)^2 + 0.9090}$$

transforms to

$$\frac{Z^2 - Z \left[e^{-0.2303T} (\cos 0.9534T) \right]}{Z^2 - Z \left[2e^{-0.2303T} (\cos 0.9534T) \right] + e^{-2(0.2303)T}} \qquad (2.21)$$

$$\frac{0.9534}{(S + 0.2303)^2 + 0.9090}$$

transforms to

$$\frac{Z \left[e^{-0.2303T} (\sin 0.9534T) \right]}{Z^2 - Z \left[2e^{-0.2303T} (\cos 0.9534T) \right] + e^{-2(0.2303)T}} \qquad (2.22)$$

Therefore

$$G(Z) \times T = 0.4606TZ \left\{ \frac{1}{Z - e^{-0.4606T}} \right.$$

$$- \frac{Z - \left[e^{-0.2303T} (\cos 0.9534T) \right]}{Z^2 - Z \left[2e^{-0.2303T} (\cos 0.9534T) \right] + e^{-0.4606T}}$$

$$+ \left. \frac{0.2416 \left[e^{-0.2303T} (\sin 0.9534T) \right]}{Z^2 - Z \left[2e^{-0.2303T} (\cos 0.9534T) \right] + e^{-0.4606T}} \right\} \qquad (2.23)$$

(c) The corresponding frequency response is obtained by substituting $e^{j\omega T}$ for Z in equation 2.23, thus giving

$$G(e^{j\omega T}) = 0.4606T \, e^{j\omega T} \left\{ \frac{1}{e^{j\omega T} - e^{-0.4606T}} \right.$$

$$-\frac{e^{j\omega T} - [e^{-0.2303T}(\cos 0.9534T)]}{e^{j2\omega T} - e^{j\omega T}[2e^{-0.2303T}(\cos 0.9534T)] + e^{-0.4606T}}$$

$$+\left.\frac{0.2416\,e^{-0.2303T}(\sin 0.9534T)}{e^{j2\omega T} - e^{j\omega T}[2e^{-0.2303T}(\cos 0.9534T)] + e^{-0.4606T}}\right\}$$

$$(2.24)$$

Using equation 2.24 the amplitude/frequency and phase/frequency character-
istics of $G(Z)$ may be determined by calculating $|G(e^{j\omega T})|$ and $\angle G(e^{j\omega T})$
respectively over a suitable normalised-frequency range, and for a particular
value of T, see figure 2.8 and figure 2.9.

(d) The impulse response, $g(t) = \mathcal{L}^{-1}[G(S)] = \mathcal{L}^{-1}$ {equation 2.19}. Using a
table of Laplace transforms we obtain

$$g(t) = 0.4606\left\{e^{-0.4606t} - [e^{-0.2303t}\cos 0.9534t]\right.$$

$$\left. + 0.2416\,[e^{-0.2303t}\sin 0.9534t]\right\} \qquad (2.25)$$

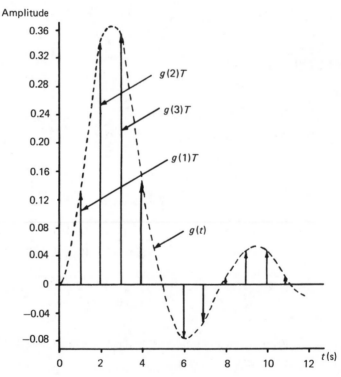

Figure 2.10   Impulse response, $g(t)$, and corresponding unit-sample response, $g(n)T$, for third-
order Chebyshev lowpass filter (example 2.5)

Figure 2.10 shows the impulse response, $g(t)$, corresponding to equation 2.25.
(e) With $T = 1$ s equation 2.20 may be represented as shown in figure 2.11a.
Referring to figure 2.11a we obtain the relationship

$$\frac{Y(Z)}{X(Z)} = \frac{1}{1 - 0.6309Z^{-1}}$$

and the corresponding linear difference equation is

$$y(n)T = x(n)T + 0.6309\, y(n-1)T \qquad (2.26)$$

Also with $T = 1$ s equation 2.21 may be represented as shown in figure 2.11b.
Referring to figure 2.11b we obtain the relationship

$$\frac{P(Z)}{X(Z)} = \frac{1 - 0.4598Z^{-1}}{1 - 0.9169Z^{-1} + 0.6309Z^{-2}}$$

and the corresponding linear difference equation is

$$p(n)T = x(n)T - 0.4598x(n-1)T + 0.9169p(n-1)T - 0.6309p(n-2)T$$
$$(2.27)$$

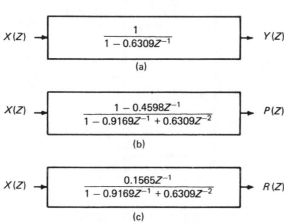

Figure 2.11 (a) Block diagram representation of equation 2.20; (b) block diagram representation of equation 2.21; (c) block diagram representation of equation 2.22

Similarly, with $T = 1$ s equation 2.22 may be represented as shown in figure
2.11c. Referring to figure 2.11c we obtain the relationship

$$\frac{R(Z)}{X(Z)} = \frac{0.1565Z^{-1}}{1 - 0.9169Z^{-1} + 0.6309Z^{-2}}$$

and the corresponding linear difference equation is

$$r(n)T = 0.1565x(n-1)T + 0.9169r(n-1)T - 0.6309r(n-2)T \qquad (2.28)$$

hence taking into account the multiplying constant: 0.4606 (see equation 2.23) the impulse-invariant digital filter derived in this example may be represented in *parallel form* as shown in figure 2.12.

To obtain the unit-sample response, $g(n)T$, we define the filter input signal as

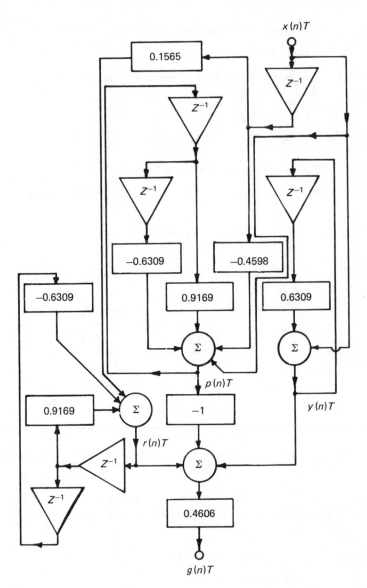

Figure 2.12 Signal flow diagram (parallel form) for third-order Chebyshev lowpass filter (example 2.5)

$$\begin{cases} x(n)T = 1 \text{ for } n = 0 \\ x(n)T = 0 \text{ for } n \neq 0 \end{cases}$$

and using figure 2.12 the filter output may be computed for integer values of $n \geqslant 0$. The resultant unit-sample response for $T = 1$ s is shown in figure 2.10.

Comments on results obtained in this example

(1) Referring to figure 2.8 and figure 2.9 we see that as T decreases the aliasing errors become insignificant and the continuous filter and digital filter frequency response characteristics have good agreement—see the characteristics shown in figure 2.8 and figure 2.9 when $T = 0.1$ s and when $T = 1.5$ s. However, when the value of T is too high, for example when $T = 1.5$ s (see figure 2.8 and figure 2.9) the response curve degenerates rapidly above $\omega \approx 2.1$; this occurs because the response repeats every $\omega_s = 2\pi/T$ rad/s.

(2) the values of the unit-sample response, $g(n)T$, are equal to the Chebyshev filter impulse response, $g(t)$, at the sampling instants. Hence we see that the derived digital filter is indeed impulse-invariant, that is, the continuous filter impulse response is maintained at each sampling instant.

(3) The derived impulse-invariant digital filter has a non-linear phase characteristic.

(4) Inspection of figure 2.12 shows that *feedback paths* exist; consequently the derived impulse-invariant digital filter is a recursive type. Furthermore, the factor equal to the sampling period, T, is absent from figure 2.12; this arises because $g(n)T$ is a time-domain description of the filter and does not require the multiplying factor, T, so that the spectra of $x(n)T$ will adequately represent $x(t)$.

Clearly, in the foregoing example the frequency response obtained depends on the sampling frequency used, and it is worth noting that in practice it may not be possible to achieve a suitable value for ω_s because its upper limit will be restricted by

(1) the sampler $(A \rightarrow D)$ operating speed;
(2) the time to execute a single *iterative-loop* of the process used for computing the filter's sampled-data output value, $y(n)T$; and
(3) the output converter $(D \rightarrow A)$ operating speed.

Consequently in designing and implementing impulse-invariant digital filters care must be taken to ensure that the value of ω_s does not produce unacceptable aliasing errors. Thus to avoid unacceptable errors the desired frequency response of $G(S)$ must be insignificant above $\omega_s/2$—this being in essence a brief statement of the sampling theorem.

If it is not possible to have a value of ω_s which avoids unacceptable aliasing errors then it will be necessary to bandlimit the continuous input signal using a highly selective guard filter, for example Butterworth or Chebyshev. However, to avoid signal distortion the guard filter should have a linear phase characteristic, which implies the use of allpass filters, and hence this is an obvious drawback in using this method of avoiding aliasing errors.

For non-bandlimited filters, such as highpass and bandstop, the impulse-invariant design method is inadequate and we must use an alternative method, such as the bilinear Z-transform, which is discussed in the next section of this chapter.

2.2.3 Bilinear Z-transform Design Method[2,3,4]

The bilinear Z-transform is characteristically bandlimiting in its action. This band-limiting property arises when the S-plane representation of the prototype continuous filter transfer function, $G(S)$, is replaced by an equivalent R-plane representation, where $S = \sigma + j\omega$ and $R = u + jv$, the relationship of the two complex variables being

$$S = \frac{2}{T} \tanh\left[\frac{RT}{2}\right]$$

where T is the sampling period. Now if we let $Z = e^{RT}$ then the relationship between S and Z is

$$S = \frac{2}{T}\left[\frac{Z-1}{Z+1}\right] \tag{2.29}$$

Thus $G(S)$ may be transformed to $G(Z)$ via the bilinear Z-transform as shown in figure 2.13. Hence it is seen that the entire left-hand half of the S-plane transforms into strips in the R-plane, which have a width equal to $2\pi/T$ rad/s (namely a bandlimited region) and furthermore it is seen that if $Z = e^{RT}$ then the strips in the left-hand half of the R-plane transform to cover the area contained inside the unit-circle in the Z-plane.

In comparison with the impulse-invariant design method the bilinear Z-transform design method is simpler to apply because it is not necessary to obtain partial-

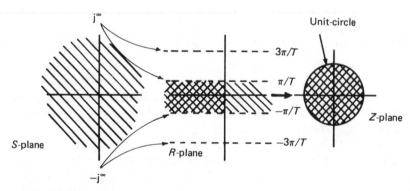

Figure 2.13 The bilinear Z-transform

fraction expansion of $G(S)$ in order to establish a transformation from the S-domain to the Z-domain using a table of Z-transforms. Indeed it is simply a matter of substituting a function of Z for each Laplace operator S, or a power of S, appearing in $G(S)$. A further advantage is that the order of the filter is unchanged because there is no need to add in an additional guard filter. However, this method does have three main inherent disadvantages, namely that a warping of the frequency scale exists due to its bandlimiting characteristic, the phase/frequency characteristic of the filter is not preserved and the frequency and time response of $G(Z)$ may differ significantly from the desired simulation of $G(S)$. The problem of warping is overcome by *pre-warping* the corner (transition) frequencies, as illustrated in example 2.6. The problem of preserving the phase/frequency characteristic may be resolved by using the guard filter approach.

The effect of warping may be seen by letting $S = j\omega_{ca}$ and $Z = e^{j\omega_{cd}T}$, where ω_{ca} and ω_{cd} refer respectively to the continuous filter and the derived digital filter. Therefore

$$j\omega_{ca} = \frac{2}{T}\left[\frac{e^{j\omega_{cd}T} - 1}{e^{j\omega_{cd}T} + 1}\right] = \frac{2}{T}\left[\frac{1 - e^{-j\omega_{cd}T}}{1 + e^{-j\omega_{cd}T}}\right]$$

$$= \frac{2}{T}\left[\frac{e^{j\omega_{cd}T/2}/j2}{e^{j\omega_{cd}T/2}/j2}\right]\left[\frac{1 - e^{-j\omega_{cd}T}}{1 + e^{-j\omega_{cd}T}}\right]$$

$$= \frac{2}{T}\left[\frac{(e^{j\omega_{cd}T/2} - e^{-j\omega_{cd}T/2})/j2}{(e^{j\omega_{cd}T/2} + e^{-j\omega_{cd}T/2})/j2}\right]$$

$$= \frac{2}{T}\left[\frac{\sin \omega_{cd}T/2}{\cos \omega_{cd}T/2/j}\right] = \frac{j2}{T}\tan \omega_{cd}T/2$$

Now pre-warping the frequency scale, the cutoff frequency of the continuous filter is given by

$$\omega_{ca} = \frac{2}{T}\tan \omega_{cd}T/2 \tag{2.30}$$

The relationship between ω_{ca} and ω_{cd} is shown in figure 2.14, and we see that the frequency scale of the digital filter is not linearly related to that of the continuous filter, that is, warping of the frequency scale exists. In order to take account of the warping, the following steps must be used when designing a digital filter using the bilinear Z-transform

(1) From the specified passband of the required digital filter and the sampling frequency, ω_{ca} is calculated using equation 2.30.
(2) $G(S)$ is chosen or derived having a response curve of the correct shape to satisfy the specification defined by the frequencies calculated in step (1).
(3) $S = 2/T\ [(Z - 1)/(Z + 1)]$ is substituted in $G(S)$, thus producing $G(Z)$.

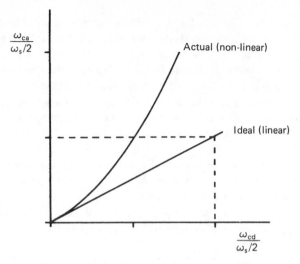

Figure 2.14 Warping of the frequency scale due to the bilinear Z-transform

Table 2.3 Frequency transformations used with indirect design methods

To transform from normalised lowpass to	Substitute for S
Lowpass	S/ω_{ca}
Highpass	ω_{ca}/S
Bandstop	$S(\omega_{cau} - \omega_{cal})/(S^2 + \omega_{cal}\omega_{cau})$
Bandpass	$(S^2 + \omega_{cal}\omega_{cau})/S(\omega_{cau} - \omega_{cal})$

ω_{cal} = lower transition frequency and ω_{cau} = upper transition frequency

The bilinear Z-transform is mainly, but not exclusively, used in the design of highpass and bandstop filters. These are commonly derived from normalised lowpass prototype filters by the application of lowpass to highpass or lowpass to bandstop transformations, see table 2.3.

The design method is illustrated in the following example.

Example 2.6

Using the bilinear Z-transform derive the digital equivalent of a second-order Butterworth lowpass filter which has the following specification

(1) digital filter cutoff frequency, $f_{cd} = 100$ Hz; and
(2) sampling period, $T = 1.6$ ms.

Obtain the amplitude and phase response for $G(S)$ and $G(Z)$.

SOLUTION

The prototype continuous filter is given by $G(S) = 1/(S^2 + (\sqrt{2})S + 1)$, see table 2.1. Substituting the numerical values in equation 2.30 yields

$$\omega_{ca} = \frac{2}{1.6 \times 10^{-3}} \tan\left[200\pi \times \frac{1.6 \times 10^{-3}}{2}\right] = 687.2 \text{ rad/s}$$

Now referring to table 2.3 it is seen that transformation from normalised lowpass to lowpass filter is achieved by substituting S/ω_{ca} for S in $G(S)$ as follows

$$G(S)_{pwt} = \frac{1}{\left[\dfrac{S}{687.2}\right]^2 + \dfrac{(\sqrt{2})S}{687.2} + 1}$$

(the pre-warped transformed transfer function)

$$G(S)_{pwt} = \frac{472243.84}{S^2 + 971.85S + 472243.84} \tag{2.31}$$

For the bilinear Z-transform

$$S = \frac{2}{T}\left[\frac{(Z-1)}{(Z+1)}\right] = 1250\left[\frac{(Z-1)}{(Z+1)}\right] \tag{2.32}$$

and it follows that

$$S^2 = 156.25 \ 10^4 \left[\frac{(Z-1)^2}{(Z+1)^2}\right] \tag{2.33}$$

Substituting equations 2.32 and 2.33 in equation 2.31 we obtain the digital filter pulse transfer function as follows

$$G(Z) = \frac{472243.84}{\left[156.25 \times 10^4 \dfrac{(Z-1)^2}{(Z+1)^2}\right] + \left[1214812.5 \dfrac{(Z-1)}{(Z+1)}\right] + 472243.84}$$

Therefore

$$G(Z) = \frac{Z^2 + 2Z + 1}{6.88 \ Z^2 - 4.62 \ Z + 1.74} \tag{2.34}$$

The frequency response of $G(S)$ is obtained by substituting $j\omega$ for S in $G(S)$, thus giving

$$G(j\omega) = \frac{1}{(1 - \omega^2) + [j(\sqrt{2})\omega]} \tag{2.35}$$

Using equation 2.35 the amplitude/frequency and phase/frequency characteristics of $G(S)$ may be determined by calculating $|G(j\omega)|$ and $\angle G(j\omega)$ respectively over

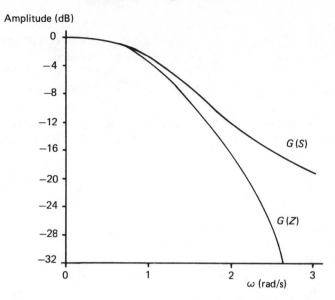

Figure 2.15   Amplitude/frequency response for second-order Butterworth lowpass filter (example 2.6)

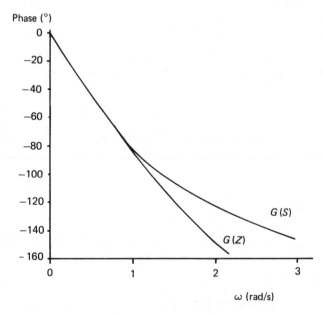

Figure 2.16   Phase/frequency response for second-order Butterworth lowpass filter (example 2.6)

a suitable normalised-frequency range, for example, $0 \leqslant \omega \leqslant 3$. These character-istics are shown in figure 2.15 and figure 2.16.

The frequency response of $G(Z)$ is obtained by substituting $e^{j\omega T}$ for Z in equation 2.34, thus giving

$$G(e^{j\omega T}) = \frac{e^{j2\omega T} + 2 e^{j\omega T} + 1}{6.88 \, e^{j2\omega T} - 4.62 \, e^{j\omega T} + 1.74}$$

$$= \frac{\cos 2\omega T + j \sin 2\omega T + 2 \cos \omega T + j2 \sin \omega T + 1}{6.88 (\cos 2\omega T + j \sin 2\omega T) - 4.62 (\cos \omega T + j \sin \omega T) + 1.74}$$

Therefore

$$G(e^{j\omega T}) = \frac{(\cos 2\omega T + 2 \cos \omega T + 1) + j(\sin 2\omega T + 2 \sin \omega T)}{(6.88 \cos 2\omega T - 4.62 \cos \omega T + 1.74) + j(6.88 \sin 2\omega T - 4.62 \sin \omega T)} \tag{2.36}$$

Using equation 2.36 the amplitude/frequency and phase/frequency characteristics of $G(Z)$ may be determined by calculating $|G(e^{j\omega T})|$ and $\angle G(e^{j\omega T})$ respectively over a suitable normalised-frequency range, for example, $0 \leqslant \omega \leqslant 3$. These characteristics are shown in figure 2.15 and figure 2.16.

2.2.4 Matched Z-transform Design Method[5,6]

The matched Z-transform directly maps the S-plane poles and zeros of $G(S)$ to corresponding poles and zeros in the Z-plane. Real poles or zeros are mapped using the relationship

$$S + \alpha \text{ transforms to } 1 - e^{-\alpha T} Z^{-1} \tag{2.37}$$

In contrast, complex poles or zeros are mapped using the relationship

$$(S + \alpha)^2 + \beta^2 \text{ transforms to } 1 - 2 e^{-\alpha T} \cos (\beta T) Z^{-1} + e^{-2\alpha T} Z^{-2} \tag{2.38}$$

The following example illustrates the method.

Example 2.7
(a) Using the matched Z-transform obtain the pulse transfer function corresponding to $G(S) = (S^2 + 2S + 5)/(S^2 + S + 1.25)$; use a value of $T = 0.1$ s.
(b) Obtain the amplitude and phase response for $G(S)$ and $G(Z)$.

SOLUTION
$G(S)$ has zeros at $S = -1 + j2$ and $S = -1 - j2$, and it has poles at $S = -\frac{1}{2} + j1$ and $S = -\frac{1}{2} - j1$. Therefore

$$G(S) = \frac{(S + 1 + j2) (S + 1 - j2)}{(S + \frac{1}{2} + j1) (S + \frac{1}{2} - j1)} = \frac{(S + 1)^2 + 4}{(S + \frac{1}{2})^2 + 1}$$

Now using equation 2.38 and $T = 0.1$ s we obtain

$$G(Z) = \frac{1 - 2\,e^{-0.1}\cos(0.2)\,Z^{-1} + e^{-0.2}\,Z^{-2}}{1 - 2\,e^{-0.05}\cos(0.1)\,Z^{-1} + e^{-0.1}\,Z^{-2}}$$

Therefore

$$G(Z) = \frac{1 - 1.7736Z^{-1} + 0.8187Z^{-2}}{1 - 1.8929Z^{-1} + 0.9048Z^{-2}} \qquad (2.39)$$

The frequency response of $G(S)$ is obtained by substituting $j\omega$ for S in $G(S)$, thus giving

$$G(j\omega) = \frac{(5 - \omega^2) - j2\omega}{(1.25 - \omega^2) - j\omega} \qquad (2.40)$$

Using equation 2.40 the amplitude/frequency and phase/frequency characteristics of $G(S)$ may be determined by calculating $|G(j\omega)|$ and $\angle G(j\omega)$ respectively over a suitable normalised-frequency range, for example, $0 \leqslant \omega \leqslant 3$. These characteristics are shown in figure 2.17 and figure 2.18.

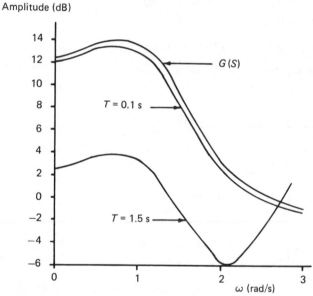

Figure 2.17   Amplitude/frequency response for example 2.7

The frequency response of $G(Z)$ is obtained by substituting $e^{-j\omega T}$ for Z^{-1} in equation 2.39, thus giving

$$G(e^{j\omega T}) = \frac{(\cos 2\,\omega T - 1.7736\cos\omega T + 0.8187) + j(\sin 2\,\omega T - 1.7736\sin\omega T)}{(\cos 2\,\omega T - 1.8929\cos\omega T + 0.9048) + j(\sin 2\,\omega T - 1.8929\sin\omega T)}$$

$$(2.41)$$

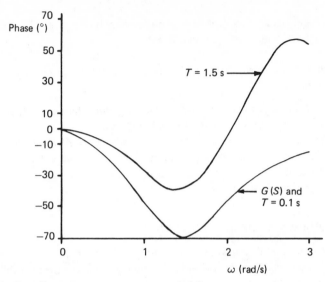

Figure 2.18 Phase/frequency response for example 2.7

Using equation 2.41 the amplitude/frequency and phase/frequency characteristics of $G(Z)$ may be determined by calculating $|G(e^{j\omega T})|$ and $\angle G(e^{j\omega T})$ respectively over a suitable normalised-frequency range, for example, $0 \leqslant \omega \leqslant 3$. These characteristics are shown in figure 2.17 and figure 2.18.

Example 2.8
Repeat example 2.7 with a new value of $T = 1.5$ s, and comment on the results.

SOLUTION
Using equation 2.38 and $T = 1.5$ s we obtain

$$G(Z) = \frac{1 - 2 e^{-1.5} \cos 3 Z^{-1} + e^{-3} Z^{-2}}{1 - 2 e^{-0.75} \cos 1.5 Z^{-1} + e^{-1.5} Z^{-2}}$$

Therefore

$$G(Z) = \frac{1 + 0.4418 Z^{-1} + 0.0498 Z^{-2}}{1 - 0.0668 Z^{-1} + 0.2231 Z^{-2}} \qquad (2.42)$$

Therefore

$$G(e^{j\omega T}) = \frac{(\cos 2\omega T + 0.4418 \cos \omega T + 0.0498) + j(\sin 2\omega T + 0.4418 \sin \omega T)}{(\cos 2\omega T - 0.0668 \cos \omega T + 0.2231) + j(\sin 2\omega T - 0.0668 \sin \omega T)} \qquad (2.43)$$

Using equation 2.43 $|G(e^{j\omega T})|$ and $\angle G(e^{j\omega T})$ were calculated for values of normalised-frequency in the range $0 \leqslant \omega \leqslant 3$. These characteristics are shown in figure 2.17 and figure 2.18. It is seen from inspection of figure 2.17 and 2.18 that when $T = 1.5$ s aliasing errors are significant.

The matched Z-transform method is easily applied, but it does have two main drawbacks, namely

(1) If $G(S)$ has zeros with centre frequencies greater than $\omega_s/2$, the corresponding zeros in $G(Z)$ will produce severe aliasing errors, and
(2) if $G(S)$ is an all-pole filter, then correspondingly $G(Z)$ will be an all-pole digital filter and will, in many cases, not adequately represent $G(S)$. However, the addition of zeros at $Z = -1$ (that is at $\omega_s/2$) is a modification sometimes used to produce satisfactory results.

2.3 FREQUENCY SAMPLING FILTERS[7,8]

Firstly it will be useful to recall that a non-repetitive time-domain waveform is related to its corresponding continuous frequency spectrum through the well known Fourier integral equations, namely

$$G(\omega) = \int_{-\infty}^{\infty} f(t) \, e^{-j\omega t} \, dt \tag{2.44}$$

and

$$f(t) = \frac{1}{2\pi} \int_{-\infty}^{\infty} G(\omega) \, e^{j\omega t} \, d\omega \tag{2.45}$$

For example, let us consider the case of a time-domain waveform defined as

$$f(t) = e^{j2\pi nt/\tau} \Big|_{-\tau/2 \leqslant t \leqslant \tau/2}$$

where n is an integer. Using equation 2.44 we obtain

$$G(\omega) = \int_{-\tau/2}^{\tau/2} (e^{j2\pi nt/\tau}) \, e^{-j\omega t} \, dt$$

and we know that $\omega = 2\pi f$. Therefore

$$G(\omega) = \int_{-\tau/2}^{\tau/2} e^{j2\pi(n/\tau - f)t} \, dt$$

$$= \left[\frac{e^{j2\pi(n/\tau - f)t}}{j2\pi(n/\tau - f)} \right]_{-\tau/2}^{\tau/2}$$

Therefore

$$G(\omega) = \frac{\tau \sin \pi \tau (f - n/\tau)}{\pi \tau (f - n/\tau)} \tag{2.46}$$

Hence we may write

$$\sum_n M_n \, e^{j2\pi nt/\tau} \; \rightleftharpoons \; \sum_n M_n \; \frac{\tau \sin \pi \tau \, (f - n/\tau)}{\pi \tau \, (f - n/\tau)} \tag{2.47}$$

where M_n is a scaling factor associated with a particular specifying number n. We see that a composite frequency response may be obtained from the summation of several individual elemental frequency responses, see equation 2.47 and figure 2.19 and figure 2.20.

In the above example the time-domain waveform has real and imaginary parts, that is

$$f(t) = e^{j2\pi nt/\tau} = \cos (2\pi nt/\tau) + j \sin (2\pi nt/\tau)$$

However, in practice for realisability the elemental filters would use a real, time-limited, elemental cosine impulse response instead of the exponential form.

To see how a real, time-limited, elemental cosine impulse response is obtained

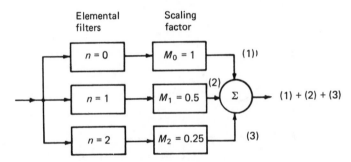

Figure 2.19 Block diagram for obtaining composite frequency response

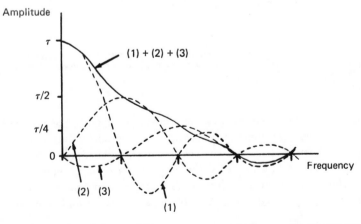

Figure 2.20 Summation of elemental frequency responses to form composite response

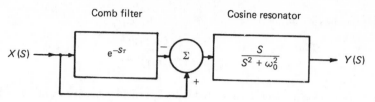

Figure 2.21 Block diagram of comb filter/cosine resonator combination

let us consider the system arrangement shown in figure 2.21, which shows a comb
filter/cosine resonator combination. The comb filter's impulse response is shown
in figure 2.22a. The cosine resonator's impulse response is

$$g(t) = \frac{2\pi nt}{\tau}\bigg|_{t \geqslant 0}$$

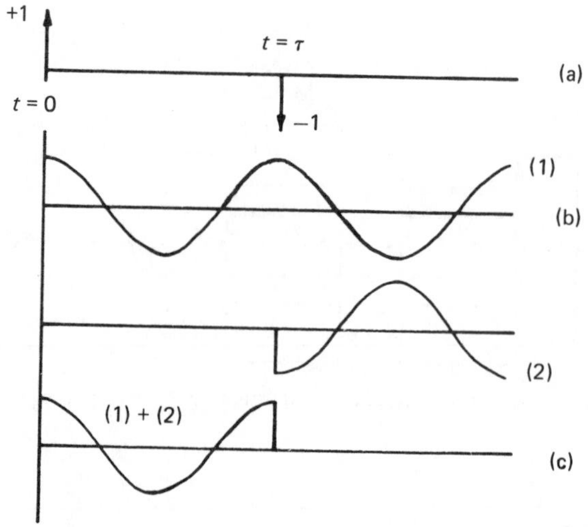

Figure 2.22 (a) Comb filter's impulse response; (b) cosine resonator's response to positive
and negative unit-impulses; (c) time limited cosine impulse response produced by comb filter/
cosine resonator combination

When the comb filter's impulse response is used as an input to the cosine resonator, the second negative impulse response produces a resonator output which is in
antiphase to the output produced by the first positive impulse (see figure 2.22b),
thereby producing a time-limited cosine impulse response, as shown in figure
2.22c. Referring to figure 2.21 (the comb filter/cosine resonator combination)
the system transfer function is $G(S) = S(1 - e^{-S\tau})/(S^2 + \omega_0^2)$, and the

corresponding amplitude/frequency and phase/frequency characteristics are specified by

$$|G(j\omega)| = \left| \frac{2\omega}{\omega^2 - \omega_0{}^2} \sin \frac{\omega\tau}{2} \right| \quad \text{and} \quad \angle G(j\omega) = -\frac{\omega\tau}{2}$$

These characteristics are shown in figure 2.23.

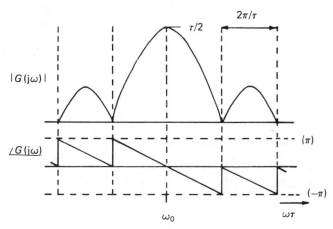

Figure 2.23 Frequency response characteristics for the comb filter/cosine resonator combination

The system shown in figure 2.21 may be expanded to include a second resonator which will obviously have its own elemental frequency response characteristics. A composite frequency response may be obtained as illustrated in figure 2.24a and figure 2.24b. In theory this type of arrangement may be expanded to include any number of resonators thereby producing the desired composite frequency response. However, it should be realised that the practical implementation of the continuous-time system shown in figure 2.21 and figure 2.24a would require the cosine resonators to be lossless, which is clearly a serious design problem. Fortunately the comb filter and cosine resonator have realisable Z-domain representations and therefore their practical implementation presents no major problems. Furthermore, digital frequency sampling filters may be implemented using a relatively simple complex number resonator. This type of resonator will be described below, and it will be used in preference to the cosine resonator. Also, we will concentrate our attention on a special class of digital frequency sampling filters whereby computational economy is achieved by multiplying sampled-data values by small integers. This type of filter has the added advantage of exhibiting linear phase characteristics. Before looking at this type of filter in detail it will be necessary to consider its two system components, namely the comb filter and the complex number resonator.

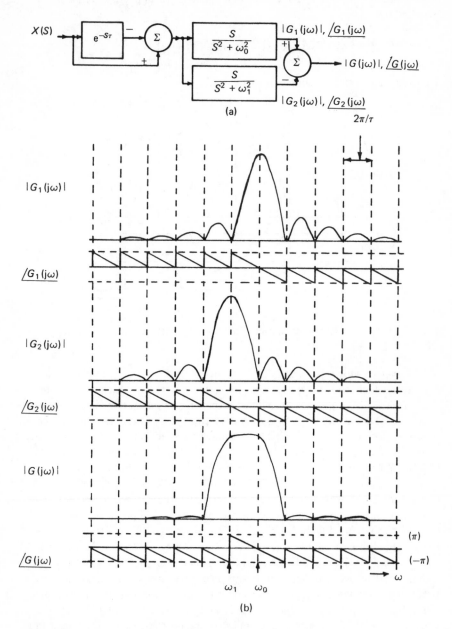

Figure 2.24 (a) System used to obtain composite frequency response; (b) frequency response characteristics of the system shown in (a)

2.3.1 Comb Filter

In figure 2.25 the comb filter's pulse transfer function is $G(Z) = 1 - Z^{-N}$, which has N zeros equally spaced around the unit-circle in the Z-plane at locations given by

$$Z_x = e^{j2\pi x/N} \quad \text{where} \quad x = 0,1,2,\ldots,(N-1) \tag{2.48}$$

and the corresponding linear difference equation for this filter is

$$y(n)T = x(n)T - x(n-N)T \tag{2.49}$$

Inspection of equation 2.49 reveals that N memory locations are required for the practical implementation of the comb filter, and correspondingly a substantial amount of memory is needed for the computational process.

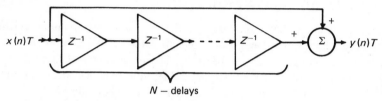

Figure 2.25   Signal flow diagram for comb filter

2.3.2 Complex Resonator

In figure 2.26, the complex resonator's pulse transfer function is $G(Z) = 1/(1 - Z^{-1} e^{j\omega_p T})$, and the pole of the resonator is used to cancel the xth zero of the comb filter. Hence the angle of the resonator pole is $\omega_p T = 2\pi x/N$. Therefore

$$G(Z) = \frac{1}{1 - Z^{-1} e^{j2\pi x/N}}$$

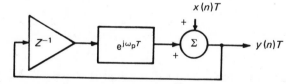

Figure 2.26   Signal flow diagram for complex resonator

2.3.3 Comb Filter/Complex Resonator Combination

From figure 2.27 we see that

$$G(Z) = G(Z)_1 \, G(Z)_2 = \frac{1 - Z^{-N}}{1 - Z^{-1} e^{j2\pi x/N}}$$

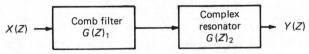

Figure 2.27   Block diagram of comb filter/complex resonator combination

For example, when using a comb filter with four delay elements, $N = 4$, and a complex resonator with a pole at $Z = 1$, to cancel the comb filter's zero at $Z = 1, x = 0$

$$G(Z) = \frac{1 - Z^{-4}}{1 - Z^{-1}} = 1 + Z^{-1} + Z^{-2} + Z^{-3} \tag{2.50}$$

The corresponding linear difference equation of this filter is

$$y(n)T = x(n)T - x(n - 4)T + y(n - 1)T \tag{2.51}$$

Note that equation 2.51 has integer coefficients.[9] The corresponding weighting function (impulse response) and pole–zero configuration for this filter are shown in figure 2.28a and figure 2.28b respectively. The amplitude/frequency and phase/frequency characteristics of the filter may be determined by substituting $e^{-j\omega T}$ for Z^{-1} in equation 2.50; these characteristics are shown in figure 2.29. Note that the nominal cutoff frequency of the filter is $\omega_c = \pi/(2T)$ rad/s, that is

$$\omega_c = \frac{2\pi}{T} \times \frac{1}{\text{number of zeros for the comb filter}} \tag{2.52}$$

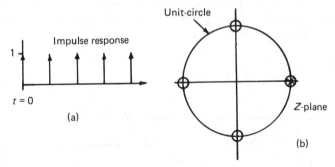

Figure 2.28   (a) Impulse response corresponding to equation 2.51; (b) Z-plane pole–zero diagram corresponding to equation 2.50

Referring to figure 2.29 and equation 2.51 we see that the filter has a lowpass characteristic, linear phase characteristic and a linear difference equation having only three terms in its right-hand side—the coefficients of these terms are not only integer but have unity value, and consequently good economy in computation is achieved.

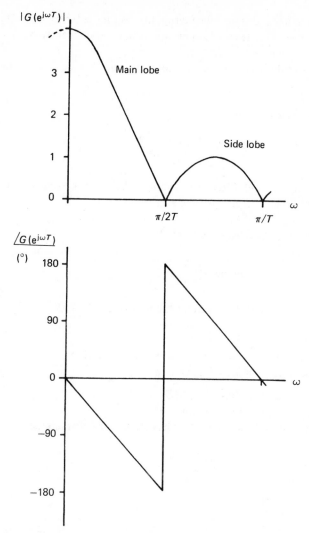

Figure 2.29 Frequency response of the filter represented by equation 2.50

The filter's magnitude/frequency characteristic has significant sidelobe levels which may be reduced by raising $G(Z)$ to an integral power; this will also make the filter's cut off much sharper.

To obtain a highpass filter characteristic the cancelling pole of the complex resonator is located at $Z = -1$. Bandpass filters may be realised by placing two cancelling poles at $Z = -j$ and $Z = +j$, yielding a passband centred on $\pi/2T$.

It has been pointed out that increasing the order of $G(Z)$ reduces sidelobe level compared with the level of the main lobe. Unfortunately the price to be paid is that the linear difference equation involves more terms in its right-hand side, which means increased computation, and consequently a loss in computational economy results.

The relationship between the mainlobe and the first sidelobe is summarised below.

(a) Lowpass or Highpass Filter

$$\text{Mainlobe: first sidelobe} = \left(N \sin \frac{3\pi}{2N} \right)^m \tag{2.53}$$

N is the number of comb filter zeros and m is the order of each zero. The gain of the filter is $(N)^m$ at the centre of the passband.

(b) Bandpass Filter Centred on $\omega = \pi/2T$

$$\text{Mainlobe: first sidelobe} = \left(\frac{N}{2} \sin \frac{3\pi}{N} \right)^m \tag{2.54}$$

The gain of the filter at the centre of the passband is $(N/2)^m$.

Example 2.9
It is required to design a highpass linear phase digital filter with a mainlobe to first sidelobe ratio of at least 5:1. Obtain:

    (a) the pulse transfer function, $G(Z)$;
    (b) the linear difference equation; and
    (c) the amplitude/frequency and phase/frequency characteristics.

SOLUTION
Using equation 2.53 and $N = 4$ yields

$$\frac{5}{1} \leqslant \left(4 \sin \frac{3\pi}{8} \right)^m$$

therefore $m = 2$ and

$$G(Z) = \frac{(1 - Z^{-4})^2}{(1 + Z^{-1})^2} = \frac{1 - 2Z^{-4} + Z^{-8}}{1 + 2Z^{-1} + Z^{-2}}$$

The corresponding linear difference equation is

$$y(n)T = x(n)T - 2x(n-4)T + x(n-8)T - 2y(n-1)T - y(n-2)T$$

The amplitude/frequency and phase/frequency characteristics are obtained by substituting $e^{-j\omega T}$ for Z^{-1} in $G(Z)$: these characteristics are shown in figure 2.30.

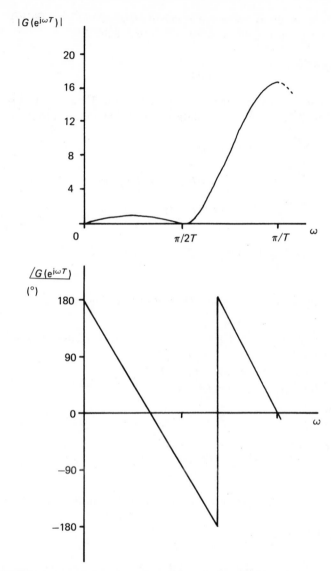

Figure 2.30 Frequency response of the filter derived in example 2.9

2.4 DIRECT APPROACH USING SQUARED MAGNITUDE FUNCTIONS[2,10]

A direct approach to the design of digital filters is to derive $G(Z)$ working in the Z-plane. When designing digital filters using this direct approach we seek functions that produce half the poles within the unit-circle, the other half being outside. These functions are known as *mirror image polynomials* (MIPs).

Now consider the magnitude squared function defined as

$$|G(e^{j\omega T})|^2 = \frac{1}{1 + [F_n(\omega T)]^2}$$

We need suitable trigonometric functions for $[F_n(\omega T)]^2$ such that on substitution $Z = e^{j\omega T}$, an MIP in the Z-plane results. One such function is $\cos^2(\omega T/2)$, that is

$$\cos^2 \frac{\omega T}{2} = \tfrac{1}{2}(1 + \cos \omega T)$$

$$= \tfrac{1}{2}[1 + \tfrac{1}{2}(e^{j\omega T} + e^{-j\omega T})]$$

$$= \tfrac{1}{2}[1 + \tfrac{1}{2}(Z + Z^{-1})]$$

Therefore

$$\cos^2 \frac{\omega T}{2} = \frac{(Z + 1)^2}{4Z}$$

It is left to the reader to verify the relationships listed below.

(1) $\sin^2(\omega T/2) = (Z - 1)^2/-4Z$ ⎫
(2) $\sec^2(\omega T/2) = 4Z/(Z + 1)^2$ ⎬ produce lowpass characteristic
(3) $\tan^2(\omega T/2) = -(Z - 1)^2/(Z + 1)^2$ ⎭
(4) $\mathrm{cosec}^2(\omega T/2) = -4Z/(Z - 1)^2$ ⎫
(5) $\cot^2(\omega T/2) = -(Z + 1)^2/(Z - 1)^2$ ⎬ produce highpass characteristic

Consider

$$|G(e^{j\omega T})|^2 = \frac{1}{1 + \left[\dfrac{\sin^2(\omega T/2)}{\sin^2(\omega_c T/2)}\right]^n}$$

where ω_c is the desired angular cutoff frequency. Substitution of $Z = e^{j\omega T}$ yields

$$|G(Z)|^2 = \frac{q^n}{q^n + p^n}$$

where $q = \sin^2(\omega_c T/2)$ and $p = (Z - 1)^2/-4Z$. The roots in the p-plane occur on a circle of radius q, thus $p_k = q\, e^{j\phi_k}$, $k = 0,1,2,\ldots,(n-1)$, and

$$\phi_k = \frac{(2k + 1)\pi}{n} \quad \text{for } n \text{ even}$$

$$\phi_k = \frac{2k\pi}{n} \quad \text{for } n \text{ odd}$$

Having solved for p we can find the corresponding factors in the Z-plane by solving $p = (Z - 1)^2/-4Z$, that is

$$-4Zp = (Z - 1)^2 = Z^2 - 2Z + 1$$

Therefore

$$Z^2 - 2Z(1 - 2p) + 1 = 0 \quad \text{(quadratic equation)}$$

Therefore

$$Z = (1 - 2p) \pm \sqrt{4p(p - 1)} \tag{2.55}$$

Hence it is seen that for every root in the p-plane there will be two corresponding roots in the Z-plane, as given by equation 2.55.

Example 2.10
A lowpass digital filter is required to have 3 dB attenuation at 1 kHz and at least 15 dB attenuation at 3 kHz. Using the direct method of design, derive $G(Z)$ to satisfy the given specification. Take $f_s = 10$ kHz.

SOLUTION

$$q = \sin^2\left(\frac{\omega_c T}{2}\right) = \tfrac{1}{2}(1 - \cos \omega_c T) = \tfrac{1}{2}[1 - \cos(2\pi \times 10^3/10^4)] = 0.0955$$

$$p = \sin^2\left(\frac{\omega T}{2}\right) = \tfrac{1}{2}(1 - \cos \omega T) = \tfrac{1}{2}[1 - \cos(2\pi \times 3 \times 10^3/10^4)] = 0.6545$$

$$|G(Z)|^2 = \frac{q^n}{q^n + p^n} = \frac{1}{1 + (p/q)^n} = \frac{1}{1 + \left[\dfrac{0.6545}{0.0955}\right]^n} = \frac{1}{1 + (6.853)^n}$$

$$15 = 10 \log_{10}[1 + (6.853)^n]$$

(see equation 2.5); solving for n we obtain $n = 1.778$, and taking the next higher integer $n = 2$. Therefore

$$k = 0 \text{ and } 1, \text{ and } \phi_0 = \frac{\pi}{2} \quad \text{and} \quad \phi_1 = \frac{3\pi}{2}$$

Therefore

$$p_0 = 0.0955 \,\underline{/\pi/2} \text{ and } p_1 = 0.0955 \,\underline{/3\pi/2}$$

Now applying equation 2.55

$$Z_0 = [1 - 2(0.0955 \,\underline{/\pi/2})] \pm \sqrt{4(0.0955 \,\underline{/\pi/2})[(0.0955 \,\underline{/\pi/2}) - 1]}$$

$$= (1 - j0.191) \pm (-0.417 + j0.459)$$

Therefore

$$Z_0 = 0.583 + j0.268 \quad \text{(inside the unit-circle)}$$

or

$$Z_0 = 1.417 - j0.649 \quad \text{(outside the unit-circle)}$$

Similarly

$$Z_1 = [1 - 2(0.0955 \angle 3\pi/2)] \pm \sqrt{4(0.0955 \angle 3\pi/2)[(0.0955 \angle 3\pi/2) - 1]}$$
$$= (1 + j0.191) \pm (0.417 + j0.459)$$

Therefore

$$Z_1 = 0.583 - j0.268 \quad \text{(inside the unit-circle)}$$

or

$$Z_1 = 1.417 + j0.649 \quad \text{(outside the unit-circle)}$$

For stability we use the poles that lie inside the unit-circle, therefore we obtain

$$G(Z) = \frac{f}{[Z - (0.583 - j0.268)] \; [Z - (0.583 + j0.268)]}$$

Now taking $|G(j\omega)| = 1$ at $\omega = 0$, then $Z = e^{j\omega T} = 1$ and therefore

$$f = 1 [1 - (0.583 - j0.268)] \; [1 - (0.583 + j0.268)] = 0.246$$

Therefore

$$G(Z) = \frac{0.246}{[(Z - (0.583 - j0.268)] \; [(Z - (0.583 + j0.268)]}$$

Table 2.4 may be used to *frequency transform* from lowpass filters to highpass, bandstop and bandpass filters. Also note that it is possible to transform from lowpass to lowpass, that is, a shift of cutoff frequency. Furthermore note that in using table 2.4 β(rad/s) is the cutoff frequency of the prototype digital filter, ω_c(rad/s) is the desired cutoff frequency, ω_1 and ω_2 (rad/s) are the lower and upper cutoff frequencies respectively and T is sampling period (s).

2.5 WAVE-SHAPING FILTERS

This type of digital filter is used to produce a certain output waveform corresponding to a specific input waveform. Thus for a given sampled-data input signal, $\{x(0)T, x(1)T, \ldots, x(n)T\}$ the corresponding digital filter output signal will be $\{y(0)T, y(1)T, \ldots, y(n)T\}$, and the filter's pulse transfer function, $G(Z)$ is

Table 2.4 Frequency transformations used with direct design methods

Filter	Substitute for Z	Design Formulae
Lowpass	$\dfrac{1 - aZ}{Z - a}$	$a = \dfrac{\sin (\beta - \omega_c)T/2}{\sin (\beta + \omega_c)T/2}$
Highpass	$-\left[\dfrac{1 + aZ}{Z + a}\right]$	$a = \dfrac{\cos (\beta - \omega_c)T/2}{\cos (\beta + \omega_c)T/2}$
Bandpass	$-\left[\dfrac{1 - \dfrac{2abZ}{(b+1)} + \dfrac{Z^2(b-1)}{(b+1)}}{\left(\dfrac{b-1}{b+1}\right) - \dfrac{2abZ}{(b+1)} + Z^2}\right]$	$a = \dfrac{\cos (\omega_2 + \omega_1)T/2}{\cos (\omega_2 - \omega_1)T/2}$ $b = \dfrac{\cot (\omega_2 - \omega_1)T/2}{1/\tan \beta T/2}$
Bandstop	$\dfrac{1 - \dfrac{2aZ}{(b+1)} + \dfrac{Z^2(1-b)}{(b+1)}}{\left(\dfrac{1-b}{b+1}\right) - \dfrac{2aZ}{(b+1)} + Z^2}$	a = same as for bandpass $b = \dfrac{\tan (\omega_2 - \omega_1)T/2}{1/\tan \beta T/2}$

$$G(Z) = \frac{\mathbf{Z}\{y(0)T, y(1)T, \ldots, y(n)T\}}{\mathbf{Z}\{x(0)T, x(1)T, \ldots, x(n)T\}} \qquad (2.56)$$

Example 2.11
The time-domain specification for a wave-shaping digital filter is

desired output waveform = $\{1, 0.25, 0.1, 0.01\}$
input waveform samples = $\{3, 1\}$

Determine the pulse transfer function, $G(Z)$, and linear difference equation of the filter.

SOLUTION
Using equation 2.56 we obtain

$$G(Z) = \frac{1 + 0.25Z^{-1} + 0.1Z^{-2} + 0.01Z^{-3}}{3 + Z^{-1}} = \frac{Y(Z)}{X(Z)} \qquad (2.57)$$

$$= \frac{Z^3 + 0.25Z^2 + 0.1Z + 0.01}{3Z^2 (Z + \frac{1}{3})} \qquad (2.58)$$

Using equation 2.57 we obtain

$$y(n)T = \tfrac{1}{3}[x(n)T + 0.25x(n-1)T + 0.1x(n-2)T + 0.01x(n-3)T - y(n-1)T]$$
(2.59)

For a recursive digital filter having M samples in the input waveform and N samples in the output waveform the corresponding linear difference equation of the filter has $M + N - 1$ terms on the right-hand side. It is possible to design a wave-shaping filter having fewer than $(M + N - 1)$ terms that will produce an acceptable output waveform approximating adequately to the desired wave-form.[11] This means that the errors introduced by dropping a term, or terms, in the linear difference equation are negligible. However, in dropping the term, or terms, it must be realised that there is a danger that the resultant filter may be unstable and therefore care must be taken to ensure that the poles of the filter lie within the unit-circle in the Z-plane.

The design of this type of filter involves two main steps

(1) finding the weights (impulse response) of the filter; and
(2) determining the corresponding filter coefficients, and thereby $G(Z)$.

The sampled input waveform is represented by the series: $I(0)T, I(1)T, I(2)T, \ldots, I(n)T$. The actual output waveform is represented by the series: $O(0)T, O(1)T, \ldots, O(m)T$. The desired output waveform is represented by the series: $O_d(0)T, O_d(1)T, \ldots, O_d(m)T$. The filter weights, $g(0)T, g(1)T, \ldots, g(m-n)T$, are convolved with the input waveform series to produce an output waveform which approximates to the specified (desired) shape. The least-squares criterion may be used requiring $[O(0)T - O_d(0)T]^2 + [O(1)T - O_d(1)T]^2 + \cdots + [O(m)T - O_d(m)T]^2$ to be minimised with respect to the weights of the filter, $g(0)T, g(1)T, \ldots, g(m-n)T$. This means that

$$F = \sum_{p=0}^{p=m} \left\{ \left[\sum_{i=0}^{i=p} I(p-i)Tg(i)T \right] - O_d(p)T \right\}^2$$
(2.60)

must be minimised with respect to $g(i)T$. That is, differentiating equation 2.60 with respect to $g(i)T$ and setting the answer to zero yields a set of simultaneous equations, from which the series of values, $g(i)T$, can be determined.

The pulse transfer function of a recursive digital filter may be represented in the form

$$G(Z) = \frac{a_0 + a_1 Z^{-1} + a_2 Z^{-2} + \ldots + a_{N-1} Z^{-(N-1)}}{1 + b_1 Z^{-1} + b_2 Z^{-2} + \ldots + b_{M-1} Z^{-(M-1)}}$$
(2.61)

the corresponding unit-impulse response being $g(0)T, g(1)T, \ldots, g(i)T$, where $i = 0, 1, 2, \ldots, (K-1)$. If $K = M + N - 1$ then a recursive realisation may be found.[12] The relationship between $G(Z)$ and $g(i)T$ may be expressed in terms of convolution of number sequences, as follows

$$\sum_{j=0}^{M-1} b_j g(k-j) \; T = \begin{cases} a_k, & k = 0,1,2, \ldots, (N-1) \\ 0, & k \geqslant N \end{cases} \qquad \text{where } b_0 = 1 \qquad (2.62)$$

Expressing equation 2.62 in matrix form we obtain

$$
\begin{bmatrix} a_0 \\ a_1 \\ a_2 \\ \cdot \\ \cdot \\ \cdot \\ a_{N-1} \\ 0 \\ \cdot \\ \cdot \\ \cdot \\ 0 \end{bmatrix}
=
\begin{bmatrix} g(0)T & 0 & 0 \ldots .0 \\ g(1)T & g(0)T & 0 \ldots .0 \\ g(2)T & g(1)T & g(0)T . . 0 \\ \cdot & & \cdot \\ \cdot & & \cdot \\ \cdot & & \cdot \\ \cdot & & \cdot \\ g(K-1)T & \ldots & g(K-M)T \end{bmatrix}
\begin{bmatrix} 1 \\ b_1 \\ b_2 \\ \cdot \\ \cdot \\ \cdot \\ b_{(M-1)T} \end{bmatrix}
\qquad (2.63)
$$

However, equation 2.63 may be partitioned, thus

$$
\begin{bmatrix} a_0 \\ \cdot \\ \cdot \\ \cdot \\ a_{(N-1)} \\ \hline 0 \\ \cdot \\ \cdot \\ \cdot \\ 0 \end{bmatrix}
=
\begin{bmatrix} g(0)T & 0 & \ldots & 0 \\ \cdot & & & \\ \cdot & & & \\ g(N-1)T & & & \\ \hline g(N)T & & & \\ \cdot & & & \\ \cdot & & & \\ g(K-1)T & \ldots & g(K-M)T \end{bmatrix}
\begin{bmatrix} 1 \\ b_1 \\ \cdot \\ \cdot \\ \cdot \\ b_{(M-1)} \end{bmatrix}
$$

That is

$$
\begin{bmatrix} a \\ \hline 0 \end{bmatrix} = \begin{bmatrix} g_1 \\ \hline g_2 \end{bmatrix} [b]
$$

Since $K = M + N - 1$ then g_2 has M columns and $M - 1$ rows, and since $b_0 = 1$ the equation for the lower partition may be written as

$$g' = -g_3 b''$$

where g' is the first column of g_2, g_3 is the remaining $M - 1$ square matrix and b'' is the b matrix with the b_0 term omitted. It follows that the b_i and a_i coefficients are found using the relationships

$$b'' = -g_3^{-1} g' \qquad (2.64)$$
$$a = g_1 b \qquad (2.65)$$

The following worked example illustrates the method.

Example 2.12
Using the time-domain specification of example 2.11 determine the pulse transfer
function, $G(Z)$ for: (a) a four-term filter; and (b) a three-term filter. Comment on
the results.

SOLUTION
(a) *Four-term filter.* To find the weights of the filter we must use equation 2.60,
that is

$$F = \left\{ [I(0)Tg(0)T - O_d(0)T]^2 + [I(1)Tg(0)T + I(0)Tg(1)T - O_d(1)T]^2 \right.$$
$$+ [I(2)Tg(0)T + I(1)Tg(1)T + I(0)Tg(2)T - O_d(2)T]^2$$
$$\left. + [I(3)Tg(0)T + I(2)Tg(1)T + I(1)Tg(2)T + I(0)Tg(3)T - O_d(3)T]^2 \right\}$$

The time-domain specification for the filter is

$$I(0)T = 3, I(1)T = 1, I(2)T = I(3)T = 0$$
$$O_d(0)T = 1, O_d(1)T = 0.25, O_d(2)T = 0.1, O_d(3)T = 0.01$$

Therefore

$$F = [10g(0)T^2 - 6.5g(0)T + 6g(0)Tg(1)T + 10g(1)T^2 - 1.7g(1)T$$
$$+ 6g(1)Tg(2)T + 10g(2)T^2 - 0.62g(2)T + 6g(2)Tg(3)T + 9g(3)T^2$$
$$- 0.06g(3)T + 1.0726] \tag{2.66}$$

We must now minimise F (equation 2.66) with respect to $g(0)T$, $g(1)T$, $g(2)T$
and $g(3)T$, that is, differentiating with respect to each of these weights and equating to zero, yielding four simultaneous equations

$$20g(0)T + 6g(1)T = 6.5$$
$$6g(0)T + 20g(1)T + 6g(2)T = 1.7$$
$$6g(1)T + 20g(2)T + 6g(3)T = 0.62$$
$$6g(2)T + 18g(3)T = 0.06$$

and solving these four equations we obtain

$$g(0)T = 0.3333, g(1)T = -0.0278, g(2)T = 0.0426, g(3)T = -0.0109$$

To find the filter coefficients we use the partitioned matrix, that is

$$
\begin{bmatrix} a_0 \\ a_1 \\ \hline 0 \\ 0 \end{bmatrix} =
\left[\begin{array}{ccc} g(0)T & 0 & 0 \\ g(1)T & g(0)T & 0 \\ \hline g(2)T & g(1)T & g(0)T \\ g(3)T & g(2)T & g(1)T \end{array} \right]
\begin{bmatrix} 1 \\ b_1 \\ \hline b_2 \end{bmatrix}
$$

Therefore equation 2.64 gives

$$\begin{bmatrix} b_1 \\ b_2 \end{bmatrix} = \begin{bmatrix} -g(1)T & -g(0)T \\ -g(2)T & -g(1)T \end{bmatrix}^{-1} \begin{bmatrix} g(2)T \\ g(3)T \end{bmatrix}$$

and solving the matrix equation $b_1 = 0.1824$ and $b_2 = -0.1126$. Now using equation 2.65 we obtain

$$\begin{bmatrix} a_0 \\ a_1 \end{bmatrix} = \begin{bmatrix} g(0)T & 0 & 0 \\ g(1)T & g(0)T & 0 \end{bmatrix} \begin{bmatrix} 1 \\ b_1 \\ b_2 \end{bmatrix}$$

and solving the matrix equation we obtain

$$a_0 = 0.3333, a_1 = 0.0330$$

Therefore

$$G(Z) = \frac{0.3333 + 0.0330Z^{-1}}{1 + 0.1824Z^{-1} - 0.1126Z^{-2}} = \frac{Y(Z)}{X(Z)} \tag{2.67}$$

$$= \frac{0.3333Z^2 + 0.0330Z}{Z^2 + 0.1824Z - 0.1126}$$

$$= \frac{0.3333Z \, (Z + 0.0990)}{(Z + 0.4389) \, (Z - 0.2565)} \tag{2.68}$$

Check: The linear difference equation corresponding to equation 2.67 is

$$y(n)T = 0.3333x(n)T + 0.0330x(n-1)T - 0.1824y(n-1) \, T$$
$$+ 0.1126y(n-2)T$$

$x(0)T = 3$ and $x(1)T = 1$ (the given input waveform samples)

$y(0)T = 0.9999$ (desired value = 1)

$y(1)T = 0.2499$ (desired value = 0.25)

$Y(2)T = 0.1$ (desired value = 0.1)

$Y(3)T = 0.0099$ (desired value = 0.01)

$y(4)T = 0.0095$ (desired value = 0)

$y(5)T = 0.0006$ (desired value = 0)

$y(6)T = 0.0012$ (desired value = 0)

$y(7)T = -0.0003$ (desired value = 0)

$y(8)T = 0.0002$ (desired value = 0)

and so on

Comments

(1) From equation 2.68 we see that the filter has poles at $Z = 0.2565$ and $Z = -0.4389$, which lie within the unit-circle in the Z-plane, and therefore the filter is stable.

(2) The actual outputs are close enough to the desired outputs to satisfy the given specification.

(b) Three-term filter. The only difference between this case and the previous case is that $g(3)T = 0$, and therefore equation 2.66 reduces to

$$F = [10g(0)T^2 - 6.5g(0)T + 6g(0)Tg(1)T + 10g(1)T^2 - 1.7g(1)T$$
$$+ 6g(1)Tg(2)T + 10g(2)T^2 - 0.62g(2)T + 1.0726] \qquad (2.69)$$

Now we minimise F (equation 2.69) with respect to $g(0)T$, $g(1)T$ and $g(2)T$, that is, differentiating with respect to these weights and equating to zero, yielding three simultaneous equations

$$20g(0)T + 6g(1)T = 6.5$$

$$6g(0)T + 20g(1)T + 6g(2)T = 1.7$$

$$6g(1)T + 20g(2)T = 0.62$$

and solving these three equations we obtain

$$g(0)T = 0.3330, g(1)T = -0.0266, g(2)T = 0.0390$$

Using equation 2.64

$$b_1 = \left\{ - [g(1)T] \right\}^{-1} \times g(2)T = \frac{0.0390}{0.0266} = 1.4662$$

Using equation 2.65

$$a_0 = g(0)T = 0.3330$$

$$a_1 = g(1)T + [g(0)T\, b_1] = -0.0266 + (0.3330 \times 1.4662)$$

Therefore

$$a_1 = 0.4616$$

Therefore

$$G(Z) = \frac{0.3330 + 0.4616Z^{-1}}{1 + 1.4662Z^{-1}}$$

$$= \frac{0.3330(Z + 1.3862)}{Z + 1.4662}$$

Hence we see that the filter has a pole at $Z = -1.4662$, which is outside the unit circle in the Z-plane, and therefore the filter is unstable. Consequently the desired output waveform will not be achieved.

References 13 and 14 provide additional information on digital filter design and digital signal processing, and both describe recursive filters.

REFERENCES

1. F. F. Kuo, *Network Analysis and Synthesis* (Wiley, New York and London, 1962) chapter 12.
2. C. M. Rader and B. Gold. 'Digital Filter Design in the Frequency Domain', *Proc. IEEE*, 55 (1967) 149–71.
3. H. J. Blinchikoff and A. I. Zverev, *Filtering in the Time and Frequency Domains* (Wiley, New York and London, 1976) chapter 9.
4. L. R. Rabiner and B. Gold, *Theory and Application of Digital Signal Processing* (Prentice-Hall, Englewood Cliffs, N.J., 1975).
5. S. A. Tretter, *Introduction to Discrete-time Signal Processing* (Wiley, New York and London, 1976).
6. J. R. Mick, *Digital Signal Processing Handbook* (Advanced Micro Devices, California, 1976).
7. R. E. Bogner and A. G. Constantinides, *Introduction to Digital Filtering* (Wiley, New York and London, 1975) chapters 8 and 9.
8. B. Gold and C. M. Rader, *Digital Processing of Signals* (McGraw-Hill, New York, 1969) chapter 3.
9. P. A. Lynn, 'Economic Linear-Phase Recursive Digital Filters' *Electron. Lett.*, 6 (1970) 143–5.
10. M. H. Ackroyd, *Digital Filters* (Butterworths, London, 1973) chapter 2.
11. J. F. Claerbout, 'Digital Filters and Applications to Seismic Detection and Discrimination', M.Sc. Thesis, M.I.T. (1963).
12. C. S. Burrus and T. W. Parks, 'Time Domain Design of Recursive Digital Filters', *Trans. Audio and Electroacoustics, IEEE*, 18 (1970) 137–41.
13. F. J. Taylor, *Digital Filter Design Handbook* (Marcel Dekker, Inc., New York, 1983)
14. M. Bellanger, *Digital Processing of Signals: Theory and Practice* (Wiley, New York and London, 1984).

PROBLEMS

2.1 The transfer function of a single section RC lowpass filter is $G(S) = \omega_0/(S + \omega_0)$, where ω_0 is the cutoff frequency (rad/s) equal to $\omega_s/4$ (ω_s being the radian sampling frequency). Using the impulse invariant design method determine the corresponding pulse transfer function, $G(Z)$. Take the sampling period, T, to be 1 s.

2.2 Using the bilinear Z-transform design method determine the pulse transfer function, $G(Z)$, corresponding to the following three filter specifications

(a) A highpass digital filter having $f_{cd} = 100$ Hz and $T = 1.6$ ms, which is based on a prototype second-order analogue Butterworth filter.
(b) A bandpass digital filter having $f_{cdl} = 80$ Hz, $f_{cdu} = 120$ Hz and $T = 1.6$ ms,

which is based on a prototype first-order analogue Butterworth filter.

(c) A bandstop digital filter having $f_{cdl} = 80$ Hz, $f_{cdu} = 120$ Hz and $T = 1.6$ ms, which is based on a prototype first-order analogue Butterworth filter.

2.3 Using the matched Z-transform design method obtain the pulse transfer function corresponding to $G(S) = (S + 0.4)/(S^2 + 0.5S + 1.0625)$. Use a value of $T = 0.5$ s.

2.4 The specification for a bandpass digital filter which is based on the comb filter/complex resonator combination is summarised as follows.

(a) The frequency response is to be centred on $\omega = \pi/2T$ by placing cancelling poles at $Z = \pm j$.

(b) The ratio of the main lobe to the first sidelobe is to be at least 4:1.

(c) The comb filter is to have eight delay elements.

Obtain

(i) The pulse transfer function, $G(Z)$, to satisfy the given specification.

(ii) The corresponding linear difference equation.

(iii) The value of the filter's unit-impulse response at the first, second and third sampling instants.

2.5 Use table 2.4 to frequency transform the digital filter pulse transfer function derived in example 2.10 to a corresponding highpass filter having a cutoff frequency of 400 Hz.

2.6 Derive the pulse transfer function, $G(Z)$, of a four-term digital wave-shaping filter which will produce an acceptable approximation to the desired output sample values: $\{2, 0.2, 0.05\}$ when the input sample values are: $\{3, 2, 1\}$.

3 Design of Non-recursive Digital Filters

3.1 INTRODUCTION

Non-recursive digital filters have a weighting sequence (impulse response), $g(i)T$, which is finite in length, and consequently this type of filter is commonly referred to as a finite impulse response (FIR) filter. The term *non-recursive* intrinsically means that the output of the filter, $y(n)T$, is computed using the present input, $x(n)T$, and previous inputs, $x(n-1)T$, $x(n-2)T$, . . ., and furthermore the filter has no inherent feedback, which means that previous output values, $y(n-1)T$, $y(n-2)T$, . . ., are not used in the computation of $y(n)T$. Figure 3.1 shows a typical example of a non-recursive filter; the arrangement shown is a digital transversal filter.[1]

A non-recursive filter has two main advantages

(1) they are always stable because: (a) there is no feedback between output and input, and (b) the impulse response is finite; and
(2) the amplitude and phase characteristics may be arbitrarily specified—the design starting off from this specification. Clearly it is possible to specify that the non-recursive digital filter has a linear phase characteristic, thereby eliminating the possibility of phase distortion in the output waveform.

The two main disadvantages of non-recursive filters are

(1) compared with a recursive counterpart, a non-recursive filter will generally use more memory and arithmetic for its implementation; and
(2) the frequency transformations used in chapter 2 are not generally suitable for the design of non-recursive filters because in applying them a recursive configuration will normally be produced. The one interesting exception is the lowpass to highpass transformation, in which the linear phase characteristic of the lowpass prototype is preserved. This transformation, lowpass to highpass, is achieved by simply changing the sign of alternate weights in the impulse response weighting sequence.

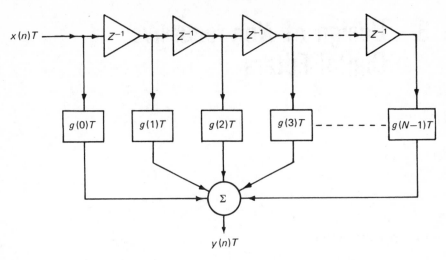

Figure 3.1 Digital transversal filter

Generally the disadvantages are easily outweighed by the advantages and consequently non-recursive digital filters are often used in practice.

3.2 FREQUENCY-DOMAIN DESIGN USING WINDOW FUNCTIONS[2,3,4,12,13]

The frequency-domain design of non-recursive digital filters using window functions is basically a frequency sampling method, and the design process involves four main steps.

(1) The ideal frequency response function (amplitude/frequency and phase/frequency characteristics) is specified. For example, the ideal amplitude/frequency and phase/frequency characteristics of a lowpass filter may be specified as shown in figure 3.2, that is, the amplitude characteristic is the ideal *brickwall* function and the phase characteristic shows zero phase shift.
(2) The specified amplitude/frequency and phase/frequency characteristics of the filter are used to compute the impulse response weighting sequence, $g(i)T$ (see equation 3.3). This step in the design process will now be examined more closely. Suppose that the ideal response function (figure 3.2) is sampled at intervals of $1/NT$ Hz along the frequency axis, then the set of frequency samples $(G_0, G_1, G_2, \ldots, G_{N-1})$ are related to the impulse response by the discrete Fourier transform (DFT) which in this case is defined as

$$G_r = \sum_{i=0}^{N-1} g(i)T \, e^{-j2\pi ir/N} \tag{3.1}$$

where $r = 0, 1, 2, \ldots, (N-1)$. Note that the G_rs are generally complex

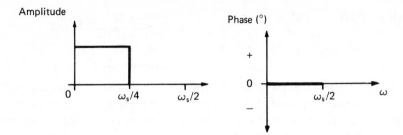

Figure 3.2 Ideal lowpass filter characteristics

numbers and that the $g(i)T$s can also be complex. For notational convenience equation 3.1 can be written as

$$G_r = \sum_{i=0}^{N-1} g(i)TW^{ir} \qquad (3.2)$$

where $W = e^{-j2\pi/N}$ and $r = 0, 1, 2, \ldots, (N-1)$. The impulse response weighting sequence $g(i)T$ may be determined by application of the inverse discrete Fourier transform (IDFT) which in this case is

$$g(i)T = \frac{1}{N} \sum_{r=0}^{N-1} G_r W^{-ir} \qquad (3.3)$$

where $i = 0, 1, 2, \ldots, (N-1)$. Referring to equation 3.2 we can write it in the form

$$G_r = g(0)TW^0 + g(1)TW^r + g(2)TW^{2r} + \ldots + g(N-2)TW^{(N-2)r} + g(N-1)TW^{(N-}$$

but $W^{Nr} = e^{-j2\pi r} = 1$ for r integer, therefore

$$G_r = g(0)TW^0 + g(1)TW^r + g(2)TW^{2r} + \ldots + g(N-2)TW^{-2r} + g(N-1)TW^{-r} \qquad (3.4)$$

Now referring to equation 3.3 we can write it in the form

$$g(i)T = \frac{1}{N}(G_0 W^0 + G_1 W^{-i} + G_2 W^{-2i} + \ldots + G_{N-2} W^{-(N-2)i} + G_{N-1} W^{-(N-1)i}$$

but $W^{-Ni} = e^{j2\pi i} = 1$ for i integer, therefore

$$g(i)T = \frac{1}{N}(G_0 W^0 + G_1 W^{-i} + G_2 W^{-2i} + \ldots + G_{N-2} W^{2i} + G_{N-1} W^i) \qquad (3.5)$$

Now for the sake of comparing the DFT and IDFT we will suppose that a computer program has been written to implement equation 3.4, then a direct comparison of equation 3.4 with equation 3.5 reveals that this same program may be used to obtain the IDFT by suitable rearrangement of the input data, namely

$$g(0)T, g(1)T, \ldots, g(N-2)T, g(N-1)T \quad : \text{input to DFT program to} \\ \text{calculate DFT}$$

$$G_0/N, G_{N-1}/N, \ldots, G_2/N, \quad G_1/N \quad : \text{input to DFT program to} \\ \text{calculate IDFT}$$

Hence it is seen that a single computer program can compute either transform, and data rearrangement is all that is needed to yield the other transformation. In practice the computation of $g(i)T$ would readily be achieved using the inverse FFT program discussed in section 1.8.

(3) The next step in the design process involves the modification of the impulse response weighting sequence, $g(i)T$. This is achieved by multiplying it with a suitable *window function*. Indeed, to reduce the *oscillations*, known as Gibbs phenomenon, which are caused by suddenly truncating the Fourier series to N values, it is appropriate to apply a suitable window (weighting function) to the truncated Fourier series. The application of the window function results in a gradual tapering of the coefficients of the series such that the middle term, $g(0)T$, is undisturbed, and the coefficients at the extreme ends of the series are negligible. Consequently the unwanted overshoot at sharp transitions in the ideal (specified) response characteristics is considerably reduced (see figure 3.3). Three window functions commonly used in the design of non-recursive digital filters are: (a) *Hamming window*, which is defined by $W_r = 0.54 + 0.46 \cos (r\pi/I)$; (b) *Blackman window*, which is defined by $W_r = 0.42 + 0.5 \cos (r\pi/I) + 0.08 \cos (2r\pi/I)$; (c) *Hanning window*, which is defined by $W_r = 0.5 + 0.5 \cos (r\pi/I)$. For each of the above window functions I is taken to be the number of

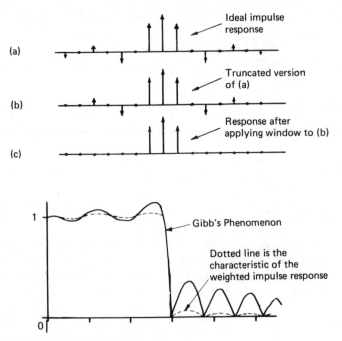

Figure 3.3 Reduction of Gibbs phenomenon

terms to be included on either side of $g(0)T$. Note that other window functions (Rectangular, Kaiser, etc.) exist, and are indeed useful. However, the three window functions listed above are generally adequate for most design problems. (4) The final step in the design process involves calculation of the filter co-efficients as $W_r g(i)T$, from which the filter's pulse transfer function, $G(Z)$, may be obtained.

The following worked example illustrates the method.

Example 3.1
Design a non-recursive digital filter based on the ideal lowpass characteristic shown in figure 3.4. Use a Hamming window and take $N = 16$ and $I = 4$. Check the design by obtaining the amplitude/frequency and phase/frequency characteristics of the derived filter.

SOLUTION
Referring to figure 3.4, the characteristic's sampled values (G_r values) are

$$(1, 1, 1, 1, 0.5, 0, 0, 0, 0, 0, 0, 0, 0.5, 1, 1, 1)$$

Note that in evaluating the response characteristics at a point of discontinuity (G_4 and G_{12}), the average of the two values immediately on the left and on the right of the discontinuity is used.

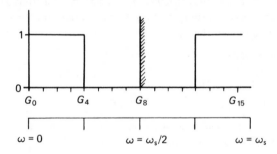

Figure 3.4 Characteristic used in example 3.1

Using equation 3.3 we obtain

$$g(i)T = \frac{1}{16} (1 + e^{+j\pi i/8} + e^{+j\pi i/4} + e^{+j\pi i/8} + 0.5e^{+j\pi i/2}$$
$$+ 0.5e^{+j3\pi i/2} + e^{+j13\pi i/8} + e^{+j7\pi i/4} + e^{+j15\pi i/8})$$

Therefore

$$g(0)T = \frac{1}{16} (1 + 1 + 1 + 1 + 0.5 + 0.5 + 1 + 1 + 1) = 0.5$$

and

$$g(1)T = \tfrac{1}{16} (1 + 0.924 - j0.383 + 0.707 - j0.707 + 0.383 - j0.924 + 0$$
$$- j0.5 + 0 + j0.5 + 0.383 + j0.924 + 0.707 + j0.707 + 0.924$$
$$+ j0.383)$$

Therefore $g(1)T = 0.314$ and similarly $g(2)T = 0, g(3)T = -0.094, g(4)T = 0$. Note that in this example $I = 4$, therefore there is no need to evaluate $g(5)T$, $g(6)T$, etc.

Now for the Hamming window $W_r = 0.54 + 0.46 \cos (r\pi/I)$, therefore

$$W_0 = 1$$
$$W_1 = 0.865$$
$$W_2 = 0.541$$
$$W_3 = 0.215$$
$$W_4 = 0.081$$

The product $g(i)TW_r$ gives: $g(0)TW_0, g(1)TW_1, g(2)TW_2, g(3)TW_3, g(4)TW_4 = (0.5, 0.272, 0, -0.02, 0)$. Therefore

$$G(Z)' = -0.02Z^3 + 0.272Z^1 + 0.5Z^0 + 0.272Z^{-1} - 0.02Z^{-3}$$

However, Z raised to a positive power implies a time advance, which in turn means that the filter would require sampled-data inputs for time $t < 0$, which clearly is impractical. To overcome this difficulty it is necessary to introduce a time shift so that the filter's pulse transfer function, $G(Z)$, contains no terms having Z raised to a positive power, that is

$$G(Z) = Z^{-3} G(Z)' \tag{3.6}$$

Therefore

$$G(Z) = -0.02 + 0.272Z^{-2} + 0.5Z^{-3} + 0.272Z^{-4} - 0.02Z^{-6} \tag{3.7}$$

The amplitude/frequency and phase/frequency characteristics, $|G(e^{j\omega T})|$ and $\angle G(e^{j\omega T})$ respectively, are obtained by substituting $e^{-j\omega T}$ for Z^{-1} in $G(Z)$: these characteristics are shown in figure 3.5.

The following comments are relevant to the design of non-recursive digital filters using window functions.

(1) Stopband attenuation can be improved by increasing the window width (that is increase I), or by using a different type of window, or by adopting both changes.

(2) Making the filter realisable by shifting the impulse response along the time-axis (see equation 3.6) has no effect on the amplitude/frequency characteristic, but it does, however, convert the zero phase characteristic of the original prototype filter to a linear phase characteristic (see figure 3.5), which for many applications is quite acceptable.

(3) The fast Fourier transform (FFT) may be used to expedite the computa-

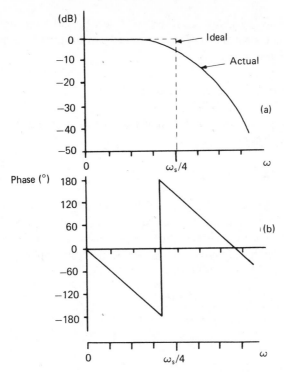

Figure 3.5(a) Amplitude/frequency response for example 3.1; (b) phase/frequency response for example 3.1

tion of the impulse response weighting sequence, $g(i)T$ (see equation 3.3 and chapter 1).

(4) The choice of N (the number of points at which the specified amplitude/frequency and phase/frequency characteristics are sampled) is not too critical.

It is desirable that N is an integral power of 2 so that it will be compatible with the FFT. Furthermore, N must be at least equal to the width of the window, that is, $N \geqslant I$. In many practical designs N is chosen so that the interval between the sampling points (for example, spacing between G_0 and G_1) is a fraction (typically $\leqslant 1/10$) of the widths of the transition bands required in the final filter design.

3.3 EQUIRIPPLE APPROXIMATION METHOD[5,6,7,8,9,10]

This design method is concerned with the zero-phase non-recursive (FIR) filter having a frequency response characteristic defined by

$$G(e^{j\omega T}) = \sum_{i=-M}^{M} g(i)T\, e^{-ji\omega T} \qquad (3.8)$$

The impulse response weighting sequence, $g(i)T$, has $N = 2M + 1$ values, and for zero-phase $g(i)T$ must equal $g(-i)T$. The symmetry of the impulse response means that equation 3.8 can be expressed as

$$G(e^{j\omega T}) = g(0)T + \sum_{i=1}^{M} 2g(i)T \cos (i\omega T) \qquad (3.9)$$

Furthermore, since $\cos (i\omega T)$ can be expressed as a sum of powers of $\cos \omega T$, equation 3.9 can be expressed as

$$G(e^{j\omega T}) = \sum_{i=0}^{M} a_i (\cos \omega T)^i \qquad (3.10)$$

where the a_is are constant coefficients which are related to the values of $g(i)T$.

An equiripple approximation to a lowpass filter is shown in figure 3.6, and we see that the approximation is required to approximate to 1 in the passband: $0 \le \omega T \le \omega_{pb}T$, with maximum error δ_1, and it must approximate to 0 in the stopband; $\omega_{sb}T \le \omega T \le \pi$, with maximum error δ_2.

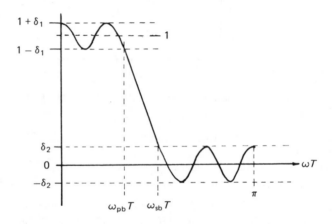

Figure 3.6 Lowpass filter equiripple approximation

In using equiripple approximations it is important to consider the number of local maxima and minima in the range $0 \le \omega T \le \pi$. Equation 3.10 is an Mth order trigonometric polynomial, and this means that there can only be $(M - 1)$ local maxima and minima in the range $0 \le \omega T \le \pi$. It is instructive to differentiate equation 3.10 with respect to ωT, as follows

$$\frac{dG(e^{j\omega T})}{d(\omega T)} = -\sin \omega T \sum_{i=1}^{M} ia_i (\cos \omega T)^{i-1} \qquad (3.11)$$

Thus it is seen from equation 3.11 that $G(e^{j\omega T})$ will be either a maximum or a minimum at $\omega T = 0$ and at $\omega T = \pi$, and consequently there will be at most $(M + 1)$ local extrema in the range $0 \le \omega T \le \pi$.

In designing FIR filters using the equiripple approximation it is not possible to specify independently each of the filter parameters $\delta_1, \delta_2, \omega_{pb}T, \omega_{sb}T$ and M. One approach developed by Hermann and Schuessler was to fix the values of δ_1, δ_2 and M, and to let $\omega_{pb}T$ and $\omega_{sb}T$ be variables. In general, there will correspondingly be N_p extrema in the passband and N_s extrema in the stopband, where

$$N_p + N_s - 1 = M \tag{3.12}$$

Hence it is possible to obtain $2(N_p + N_s - 1)$ equations relating the $(N_p + N_s)$ filter coefficients and the $(N_p + N_s - 2)$ frequencies at which extrema occur (extrema also occur at $\omega T = 0$ and $\omega T = \pi$). Unfortunately the $2(N_p + N_s - 1)$ equations are non-linear and must be solved by an iterative process. The Newton–Raphson method for solving these equations is summarised below.

The set of non-linear equations can be expressed as

$$f_j(x_1, x_2, \ldots, x_n) = 0$$

where $j = 1, 2 \ldots, n$. Taking $x_n(i)$ to be the current value of x_n, and $x_n(i + 1)$ to be the new value of x_n obtained after an iteration, the relationships between the current values of x and the new values of x are

$$
\begin{bmatrix} x_1(i+1) \\ x_2(i+1) \\ \vdots \\ \\ x_n(i+1) \end{bmatrix} = \begin{bmatrix} x_1(i) \\ x_2(i) \\ \vdots \\ \\ x_n(i) \end{bmatrix} + \left\{ \begin{bmatrix} df_1(i)/dx_1 & df_1(i)/dx_2 \ldots df_1(i)/dx_n \\ df_2(i)/dx_1 & df_2(i)/dx_2 \ldots df_2(i)/dx_n \\ \vdots & \vdots & \vdots \\ \\ df_n(i)/dx_1 & df_n(i)/dx_2 & df_n(i)/dx_n \end{bmatrix}^{-1} \begin{bmatrix} -f_1(i) \\ -f_2(i) \\ \vdots \\ \\ -f_n(i) \end{bmatrix} \right\}
$$

$$\tag{3.13}$$

The following worked example illustrates the design method.

Example 3.2
Design the non-recursive equiripple filter specified by figure 3.7. Check the design by obtaining the amplitude/frequency and phase/frequency characteristics of the derived filter.

SOLUTION
 Number of extrema in passband, $N_p = 2$
 Number of extrema in stopband, $N_s = 1$

Using equation 3.12, $M = 2$, hence four non-linear equations relating the three filter coefficients and the single frequency $(\omega_1 T)$ must be found. From inspection of figure 3.7 we see that

$$G(e^{j0}) = 1 - \delta_1 \tag{3.14}$$

$$G(e^{j\omega_1 T}) = 1 + \delta_1 \tag{3.15}$$

$$G(e^{j\pi}) = -\delta_2 \tag{3.16}$$

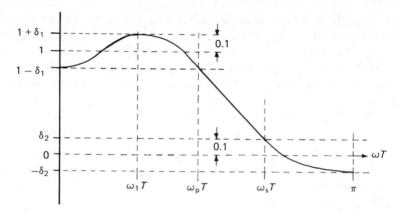

Figure 3.7 Characteristic used in example 3.2

It follows that

$$\frac{dG(e^{j\omega_1 T})}{d(\omega_1 T)} = 0 \tag{3.17}$$

For $M = 2$ equation 3.10 is

$$G(e^{j\omega T}) = a_0 + a_1 \cos \omega T + a_2 \cos^2 \omega T$$

$$= a_0 + a_1 \cos \omega T + \frac{a_2}{2}(1 + \cos 2\omega T) \tag{3.18}$$

For $M = 2$ equation 3.11 is

$$\frac{dG(e^{j\omega T})}{d(\omega T)} = -\sin \omega T (a_1 + 2a_2 \cos \omega T) \tag{3.19}$$

Thus using equations 3.14 to 3.19 inclusive, we obtain a set of four non-linear equations, namely

$$f_1(i) = a_0(i) + a_1(i) + a_2(i) + \delta_1 - 1 \tag{3.20}$$

$$f_2(i) = a_0(i) + a_1(i) \cos \omega_1 T(i) + \tfrac{1}{2} \{a_2(i)[1 + \cos 2\omega_1 T(i)]\} - \delta_1 - 1 \tag{3.21}$$

$$f_3(i) = a_1(i) + 2a_2(i) \cos \omega_1 T(i) \tag{3.22}$$

$$f_4(i) = a_0(i) - a_1(i) + a_2(i) + \delta_2 \tag{3.23}$$

Now we have to differentiate each of the four non-linear equations with respect to $a_0(i), a_1(i), a_2(i)$ and $\omega_1 T(i)$ (see equation 3.13), as follows

$$\frac{df_1(i)}{da_0(i)} = 1 \qquad \frac{df_1(i)}{da_1(i)} = 1 \qquad \frac{df_1(i)}{da_2(i)} = 1 \qquad \frac{df_1(i)}{d\omega_1 T(i)} = 0$$

$$\frac{df_2(i)}{da_0(i)} = 1 \qquad \frac{df_2(i)}{da_1(i)} = \cos \omega_1 T(i) \qquad \frac{df_2(i)}{da_2(i)} = \tfrac{1}{2}\left[1 + \cos 2\omega_1 T(i)\right]$$

$$\frac{df_2(i)}{d\omega_1 T(i)} = -\left[(a_1(i) \sin \omega_1 T(i) + a_2(i) \sin 2\omega_1 T(i)\right]$$

$$\frac{df_3(i)}{da_0(i)} = 0 \qquad \frac{df_3(i)}{da_1(i)} = 1 \qquad \frac{df_3(i)}{da_2(i)} = 2 \cos \omega_1 T(i)$$

$$\frac{df_3(i)}{d\omega_1 T(i)} = -2a_2(i) \sin \omega_1 T(i)$$

$$\frac{df_4(i)}{da_0(i)} = 1 \qquad \frac{df_4(i)}{da_1(i)} = -1 \qquad \frac{df_4(i)}{da_2(i)} = 1 \qquad \frac{df_4(i)}{d\omega_1 T(i)} = 0$$

Now taking $\delta_1 = \delta_2 = 0.1$ and $\omega_1 T(0) = 0.5$, the iterative computations (equation 3.13) produce final values

$$
\begin{aligned}
a_0 &= 0.9949 \\
a_1 &= 0.5000 \\
a_2 &= -0.5949 \\
\omega_1 T &= 1.1371
\end{aligned}
$$

Substituting these final values in equation 3.18 yields

$$G(e^{j\omega T}) = 0.6975 + 0.5 \cos \omega T - 0.2975 \cos 2\omega T$$

$$= 0.6975 + 0.25(e^{j\omega T} + e^{-j\omega T}) - 0.1488(e^{j2\omega T} + e^{-j2\omega T})$$

But $e^{j\omega T} = Z$, therefore

$$G(Z)' = 0.6975 + 0.25(Z^1 + Z^{-1}) - 0.1488(Z^2 + Z^{-2})$$

However, to make the filter causal we simply delay $g(i)T$ by M samples, that is

$$G(Z) = G(Z)' \times Z^{-2}$$

and therefore the pulse transfer function of the filter is

$$G(Z) = 0.6975Z^{-2} + 0.25Z^{-1} + 0.25Z^{-3} - 0.1488 - 0.1488Z^{-4} \qquad (3.24)$$

The amplitude/frequency and phase/frequency characteristics, $|G(e^{j\omega T})|$ and $\angle G(e^{j\omega T})$ respectively, are obtained by substituting $e^{-j\omega T}$ for Z^{-1} in $G(Z)$; these characteristics are shown in figure 3.8.

Clearly the iterative computations involved in this method are greatly facilitated with the aid of a computer program, and indeed even for relatively low values of M (say of the order of 10) computer-aided design is desirable. However, non-iterative analytic techniques for designing FIR filters have been developed; one method is described in the following section.

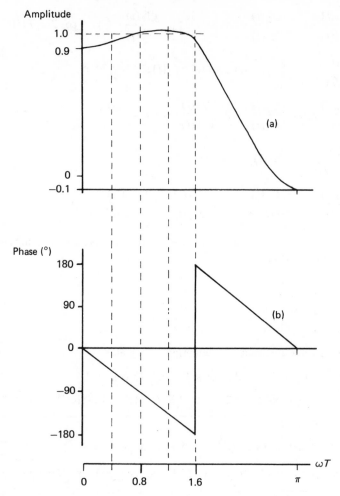

Figure 3.8(a)   Amplitude/frequency response for example 3.2; (b) phase/frequency response for example 3.2

3.4 AN ANALYTICAL TECHNIQUE FOR DESIGNING FIR FILTERS[11]

The frequency response characteristic of a zero-phase non-recursive (FIR) filter can be defined by

$$G(e^{j\omega T}) = g(0)T + \sum_{i=1}^{M} 2g(i)T \cos (i\omega T) \qquad (3.25)$$

(see equation 3.9). However, $G(e^{j\omega T})$ may be transformed into a polynomial form using

$$p = \frac{1}{2}(1 - \cos \omega T) \tag{3.26}$$

That is

$$G(e^{j\omega T}) = \sum_{i=0}^{M} Y_i p^i \tag{3.27}$$

The design of the filter relies upon finding the set of Y_i coefficients that make $G(e^{j\omega T})$ have the desired passband characteristic in the range $0 \leqslant p \leqslant p_1$ and the desired stopband characteristic in the range $p_1 \leqslant p \leqslant 1$ (see figure 3.9).

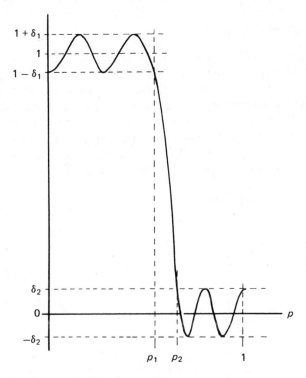

Figure 3.9 Characteristic constraints for polynomial approximations

This analytic technique uses two orthogonal polynomials to obtain the overall frequency response characteristic of the filter. The orthogonal polynomials concerned are

(1) a lowpass polynomial of the form

$$K(q) = \sum_{i=0}^{m} A_i q^i \tag{3.28}$$

where m is the passband order. This polynomial is transformed by a second polynomial
(2)

$$L(p) = \sum_{i=0}^{n} B_i p^i$$

(3.29)

where n is the order of the second polynomial. The desired overall frequency response is given by

$$G(e^{j\omega T}) = K(q)|_{q=L(p)} \quad 0 \leqslant p \leqslant 1$$

(3.30)

3.4.1 The Passband Polynomial

It is desired that the orthogonal polynomial, when normalised, must possess a cyclic behaviour with amplitude limits of ± 1 in the range $0 \leqslant q \leqslant 1$, which rapidly increases in magnitude outside this range. One suitable polynomial is the Chebyshev type, $C_n(\omega)$ (see equation 2.7 and table 2.2). A typical example is a fourth-order Chebyshev polynomial, which is shown in figure 3.10. However, it is necessary to modify the Chebyshev polynomial to make it conform to the constraints specified in figure 3.9. The modifying steps are depicted in figure 3.11, showing how the resultant function is obtained, that is

$$K(q) = 1 - \delta_1 C_n \left(\frac{q}{q_1}\right)$$

(3.31)

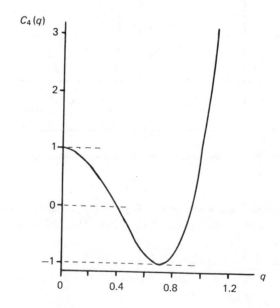

Figure 3.10 Fourth-order Chebyshev polynomial

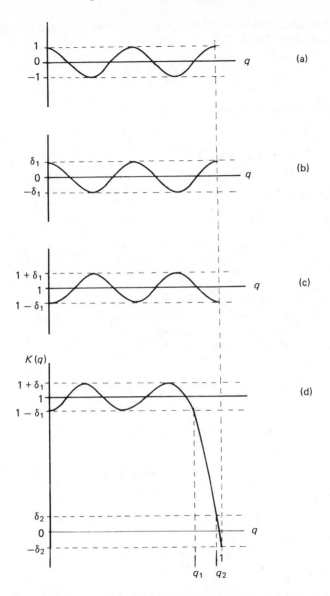

Figure 3.11 Modifying steps involved in deriving $K(q)$: (a) Chebyshev polynomial; (b) scaled Chebyshev polynomial; (c) subtract scaled Chebyshev polynomial from unity; and (d) introduction of horizontal scaling factor, q_1

3.4.2 The Overall Polynomial

The second (transforming) polynomial is orthogonal, and it may be implemented using a Chebyshev polynomial, $C_n(\omega)$. Furthermore, it is necessary to modify this normalised orthogonal polynomial to derive the transformation function, $L(p)$.

The modifying steps are depicted in figure 3.12, showing how the resultant function is obtained, that is

$$L(p) = 1 - \frac{(1 - q_2)}{2} \left\{ 1 - C_n \left[\frac{1 - p}{1 - p_2} \right] \right\}$$

(3.32)

The overall frequency response is

$$K(q)|_{q=L(p)}$$

(see equation 3.30 and figure 3.9).

The order of the filter in the p-plane is

$$M = m \times n$$

(3.33)

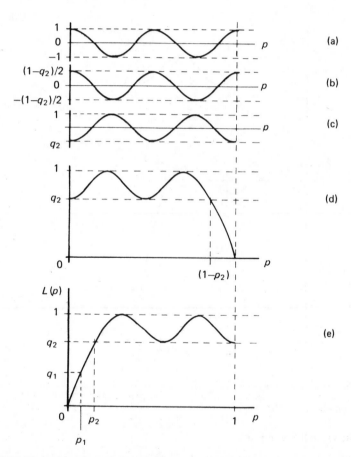

Figure 3.12 Modifying steps involved in deriving $L(p)$: (a) Chebyshev polynomial; (b) scaled Chebyshev polynomial; (c) subtract scaled Chebyshev polynomial from $[(1 - q_2)/2]$; (d) introduction of horizontal scaling factor $(1 - p_2)$; and (e) replace p by $(1 - p)$

and the corresponding Z-plane order is $2M$, that is

$$Z\text{-plane order} = 2 \times m \times n \tag{3.34}$$

3.4.3 The Design Equations

The polynomials of equations 3.31 and 3.32 must be constrained as depicted in figure 3.11d and figure 3.12e respectively, to yield the desired response shown in figure 3.9. By considering these constraints a number of design equations can be derived, as follows.

By inspecting figure 3.11d we see that at $q = 1$, $K(1) = -\delta_2$, and under this constraint equation 3.31 gives

$$-\delta_2 = 1 - \delta_1\, C_n \left(\frac{1}{q_1} \right)$$

Therefore

$$\delta_2 = \delta_1\, C_n \left(\frac{1}{q_1} \right) - 1 \tag{3.35}$$

Furthermore, from figure 3.11d we also see that at $q = q_2$, $K(q_2) = \delta_2$, and under this constraint equation 3.31 gives

$$\delta_2 = 1 - \delta_1\, C_n \left(\frac{q_2}{q_1} \right) \tag{3.36}$$

Now equating 3.35 and 3.36 we obtain

$$\delta_1 = \frac{2}{C_n(1/q_1) + C_n(q_2/q_1)} \tag{3.37}$$

By inspecting figure 3.12e we see that at $p = 0$, $L(0) = 0$. Under this constraint equation 3.32 gives

$$q_2 = 1 - \frac{2}{1 + C_n\, [1/(1 - p_2)]} \tag{3.38}$$

Also from figure 3.12e we see that at $p = p_1$, $L(p_1) = q_1$, and under this constraint equation 3.32 gives

$$q_1 = 1 - \frac{(1 - q_2)}{2} \left\{ 1 + C_n \left[\frac{1 - p_1}{1 - p_2} \right] \right\} \tag{3.39}$$

Assuming that the variables m, n, p_1 (passband cutoff frequency) and p_2 (stopband frequency) are known or given, then using the equations corresponding to the constrained conditions, δ_1, δ_2, q_1 and q_2 may be calculated for any specified orthogonal polynomial, $C_n(\omega)$. These values are then used with equations 3.31

and 3.32 to derive equation 3.30. Then equation 3.30 may be compared with equation 3.27 to yield the Y_i coefficients, and then a comparison of equations 3.27 and 3.25 yields the filter coefficients $g(0)T$ and $g(i)T$, from which the filter's pulse transfer function, $G(Z)$, is derived. The following worked example illustrates the method.

Example 3.3
Using the analytical design technique, design a non-recursive digital filter based on second-order Chebyshev polynomials. The passband and stopband cut off frequencies correspond to $p_1 = 0.3$ and $p_2 = 0.6$ respectively.

SOLUTION
For second-order Chebyshev polynomials $m = n = 2$, and therefore $M = 4$. The polynomial used in this solution is $C_n(p) = (2p^2 - 1)$, see table 2.2. Using equation 3.38 we obtain

$$q_2 = 1 - \cfrac{2}{1 + 2\left[\cfrac{1}{1 - p_2}\right]^2 - 1} = 0.84$$

Using equation 3.39 we obtain

$$q_1 = 1 - \frac{(1 - q_2)}{2}\left[1 + 2\left(\frac{1 - p_1}{1 - p_2}\right)^2 - 1\right] = 0.51$$

Using equation 3.37 we obtain

$$\delta_1 = \cfrac{2}{\left[2\left(\cfrac{1}{q_1}\right)^2 - 1\right] + \left[2\left(\cfrac{q_2}{q_1}\right)^2 - 1\right]} = 0.18$$

Using equation 3.36 we obtain

$$\delta_2 = 1 - \delta_1\left[2\left(\frac{q_2}{q_1}\right)^2 - 1\right] = 0.203$$

Using equation 3.32 we obtain

$$L(p) = 1 - \frac{(1 - q_2)}{2}\left\{1 - \left[\frac{1 - (2p^2 - 1)}{1 - p_2}\right]\right\}$$

Therefore

$$L(p) = 0.52 + 0.4p^2 \tag{3.40}$$

Using equation 3.31 we obtain

$$K(q) = 1 - \delta_1 \left[2 \left(\frac{q}{q_1} \right)^2 - 1 \right]$$

Therefore

$$K(q) = 1.18 - 0.705q^2 \qquad (3.41)$$

Using equation 3.30 we obtain

$$G(e^{j\omega T}) = 1.18 - 0.705\,(0.52 + 0.4p^2)^2$$
$$= 0.989 - 0.294p^2 - 0.113p^4 \qquad (3.42)$$

Using equation 3.27 we obtain

$$G(e^{j\omega T}) = Y_0 + Y_1 p + Y_2 p^2 + Y_3 p^3 + Y_4 p^4 \qquad (3.43)$$

Now comparing equations 3.42 and 3.43 we deduce that

$$Y_0 = 0.989$$
$$Y_1 = 0$$
$$Y_2 = -0.294$$
$$Y_3 = 0$$
$$Y_4 = -0.113$$

Using equation 3.25 we obtain

$$G(e^{j\omega T}) = g(0)T + 2\,[g(1)T \cos \omega T + g(2)T \cos 2\omega T + g(3)T \cos 3\omega T$$
$$+ g(4)T \cos 4\omega T] \qquad (3.44)$$

From equation 3.26 we see that

$$\cos \omega T = 1 - 2p$$
$$\cos 2\omega T = 2 \cos^2 \omega T - 1 = 1 - 8p + 8p^2$$
$$\cos 3\omega T = \cos \omega T\,(2 \cos 2\omega T - 1) = 1 - 18p + 48p^2 - 32p^3$$
$$\cos 4\omega T = (2 \cos^2 2\omega T) - 1 = 1 - 32p + 160p^2 - 256p^3 + 128p^4$$

Substituting the above relationships in equation 3.44 we obtain

$$G(e^{j\omega T}) = \{[g(0)T + 2g(1)T + 2g(2)T + 2g(3)T + 2g(4)T]\,p^0$$
$$- [4g(1)T + 16g(2)T + 36g(3)T + 64g(4)T]\,p^1$$
$$+ [16g(2)T + 96g(3)T + 320g(4)T]\,p^2$$
$$- [64g(3)T + 512g(4)T]\,p^3 + 256g(4)T\,p^4\} \qquad (3.45)$$

Now equating the coefficients of equations 3.43 and 3.45 we obtain the following simultaneous equations

$$0.989 = g(0)T + 2g(1)T + 2g(2)T + 2g(3)T + 2g(4)T$$

$$0 = 4g(1)T + 16g(2)T + 36g(3)T + 64g(4)T$$

$$-0.294 = 16g(2)T + 96g(3)T + 320g(4)T$$

$$0 = 64g(3)T + 512g(4)T$$

$$-0.113 = 256g(4)T$$

Solving the above equations we obtain

$g(0)T = 0.848$
$g(1)T = 0.098$
$g(2)T = -0.031$
$g(3)T = 0.004$
$g(4)T = -0.0004$

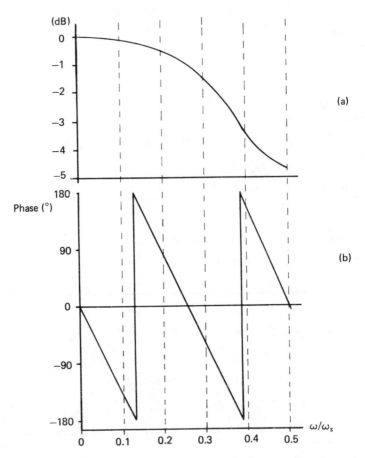

Figure 3.13(a) Amplitude/frequency response for example 3.3; (b) phase/frequency response for example 3.3

Therefore using equation 3.44 we have, for $2 \cos n\omega T = (Z^n + Z^{-n})$

$$G(Z)' = [0.848 + 0.098(Z^1 + Z^{-1}) - 0.031(Z^2 + Z^{-2})$$
$$+ 0.004(Z^3 + Z^{-3}) - 0.0004(Z^4 + Z^{-4})]$$

and to make the filter causal we multiply $G(Z)'$ by Z^{-4}, thus giving

$$G(Z) = (-0.0004 + 0.004Z^{-1} - 0.031Z^{-2} + 0.098Z^{-3}$$
$$+ 0.848Z^{-4} + 0.098Z^{-5} - 0.031Z^{-6} + 0.004Z^{-7}$$
$$- 0.0004Z^{-8})$$

The amplitude/frequency and phase/frequency characteristics, $|G(e^{j\omega T})|$ and $\angle G(e^{j\omega T})$ respectively, are obtained by substituting $e^{-j\omega T}$ for Z^{-1} in $G(Z)$; these characteristics are shown in figure 3.13.

3.5 TIME-DOMAIN DESIGN OF THE DIGITAL TRANSVERSAL FILTER

The impulse response weighting sequence $g(i)T$ values are the coefficients of the digital transversal filter shown in figure 3.1. If the non-recursive filter has to satisfy a time-domain specification the $g(i)T$ values may be determined using the least-squares criterion described in section 2.5. Referring to example 2.11 we see that the time-domain specification for a particular filter is

   desired output waveform = $\{1, 0.25, 0.1, 0.01\}$

   input waveform samples = $\{3, 1\}$

and in example 2.12 the corresponding impulse response weighting sequence of a four-term filter was determined, thus yielding

$$g(0)T = 0.3333, g(1)T = -0.0278, g(2)T = 0.0426, g(3)T = -0.0109$$

The corresponding pulse transfer function of the non-recursive filter is

$$G(Z) = 0.3333 - 0.0278Z^{-1} + 0.0426Z^{-2} - 0.0109Z^{-3} = \frac{Y(Z)}{X(Z)} \qquad (3.46)$$

Therefore the linear difference equation of this filter is

$$y(n)T = [0.3333\, x(n)T - 0.0278\, x(n-1)T + 0.0426\, x(n-2)T$$
$$- 0.0109\, x(n-3)T] \qquad (3.47)$$

Using equation 3.47 the filter output is

   $y(0)T = 0.9999$ (desired value = 1)

   $y(1)T = 0.2499$ (desired value = 0.25)

   $y(2)T = 0.1$ (desired value = 0.1)

$y(3)T = 0.0099$ (desired value = 0.01)

$y(4)T = -0.0109$ (desired value = 0)

$y(5)T = 0$ (desired value = 0)

Hence we see that the actual output waveform approximates closely to the desired output waveform.

REFERENCES

1. R. E. Bogner and A. G. Constantinides, *Introduction to Digital Filtering* (Wiley, New York and London, 1975) chapter 6.
2. B. Gold and C. M. Rader, *Digital Processing of Signals* (McGraw-Hill, New York, 1969) chapter 3.
3. L. R. Rabiner, 'Techniques for Designing Finite-Duration Impulse-Response Digital Filters', *Trans. Communication Technology, IEEE*, 19 (1971) 188-95.
4. M. H. Ackroyd, *Digital Filters* (Butterworths, London, 1973) chapter 3.
5. O. Herrmann, 'On the Design of Nonrecursive Digital Filters with Linear Phase', *Electronic Lett.*, 6 (1970) 328-9.
6. O. Herrmann and H. W. Schuessler, 'Design of Nonrecursive Digital Filters', *Electronic Lett.*, 6 (1970) 329-30.
7. T. W. Parks and J. H. McClellan, 'Chebyshev Approximation for Nonrecursive Digital Filters with Linear Phase', *Trans. Circuit Theory, IEEE*, 19 (1972) 189-94.
8. T. W. Parks and J. H. McClellan, 'A Program for the Design of Linear Phase Finite Impulse Response Filters', *Trans. Audio and Electroacoustics, IEEE*, 20 (1972) 195-9.
9. L. R. Rabiner, 'The Design of Finite Impulse Response Digital Filters Using Linear Programming Techniques', *Bell Syst. tech. J.*, (1972) 1177-98.
10. L. R. Rabiner, 'Linear Program Design of Finite Impulse Response (FIR) Digital Filters', *Trans. Audio and Electroacoustics, IEEE*, 20 (1972) 280-8.
11. J. Attikiouzel and R. Bennett, 'Analytic Techniques for Designing Digital Nonrecursive Filters', *Int. J. Electl Engng Educ.*, 14 (1977) 251-67.
12. R. W. Hamming, *Digital Filters, Second Edition* (Prentice-Hall, Englewood Cliffs, N.J., 1983).
13. C. S. Williams, *Designing Digital Filters* (Prentice-Hall, Englewood Cliffs, N.J., 1986).

PROBLEMS

3.1 Design a non-recursive digital filter based on the lowpass characteristic shown in figure 3.14. Use a Blackman window and take $N = 16$ and $I = 4$.

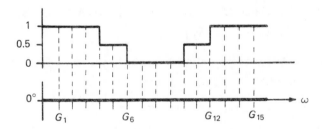

Figure 3.14   Characteristic for problem 3.1

3.2 Using the filter specification shown in figure 3.15 derive four non-linear equations to be solved using the Newton–Raphson iterative process defined in equation 3.13.

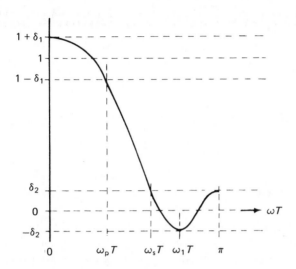

Figure 3.15   Characteristic for problem 3.2

3.3 Using the analytic design method, design a non-recursive filter based on a first-order Chebyshev polynomial. The passband and stopband cutoff frequencies correspond to $p_1 = 0.2$ and $p_2 = 0.5$ respectively.

3.4 The time-domain specification for a non-recursive digital filter is

   desired output = {0.5, 0.4, 0.2, 0.1}

   input waveform samples = {2, 0.8}

Design a four-term filter using the least-squares criterion.

4 Quantisation Considerations in Digital Filter Implementation

4.1 INTRODUCTION

Digital filters have been successfully implemented using digital minicomputers[1,2], microprocessors[3,4] and dedicated hardware.[5] The implementation of the filter involves, to some degree, using hardware having finite word lengths (16 bits, 12 bits or 8 bits are typical). These finite word lengths represent sampled-data input and output signals, filter coefficients and results of arithmetic computations; consequently inherent errors exist in the representation of these parameters, and they normally take the form of

(1) quantisation errors which arise as a result of arithmetic computations (round off and truncation errors);
(2) quantisation errors due to representing the input signal by a set of discrete values;
(3) quantisation errors due to representing the filter coefficients by a finite number of bits;
(4) limit cycle oscillations; and
(5) overflow oscillations.

The design of the filter can be undertaken without detailed consideration of the effects of finite word lengths (see chapter 2 and chapter 3). However, in contrast, in the practical implementation of the filter the errors listed above in (1) to (5) inclusive, must be considered, and they must be eliminated or kept as small as possible.

By understanding the difficulties encountered in using limited word lengths the filter designer can estimate the minimum number of bits required to form a word length in the filter implementation. Furthermore, it is possible for the designer to decide on the various trade-offs between cost and precision, thereby arriving at a satisfactory, economical and practical filter implementation.

4.2 BINARY NUMBER REPRESENTATIONS

In general any number can be represented in the form

$$N = C_y r^y + C_{y-1} r^{y-1} + \ldots + C_1 r^1 + C_0 r^0 + C_{-1} r^{-1} + \ldots \qquad (4.1)$$

where C_y is the yth coefficient and r is the radix.

The binary number system has a radix of 2, and its coefficients can only have one of two possible states, namely 0 or 1. For example, $(101*01)_2$ is a shorthand way of writing

$$1 \times 2^2 + 0 \times 2^1 + 1 \times 2^0 + 0 \times 2^{-1} + 1 \times 2^{-2}$$

which is equivalent to the number $(5.25)_{10}$. Note that the subscript after the right-hand bracket signifies the *radix* of the number. Also note that in the above binary number the symbol * is used to denote the location of the binary point, that is, the point that separates the integer part of the number from the fractional part. In the above example the binary number is represented by five binary digits (bits), three before the binary point and two after it. This number is commonly referred to as a 5-bit binary number, or a number having a word length of 5 bits. The sign of the number is represented by the leading binary digit; a positive number has 0 as the leading binary digit, and a negative number has 1 as the leading binary digit.

In digital filters two methods of implementing the arithmetic operations are encountered, namely fixed-point or floating-point arithmetic.[6] These forms are discussed below.

4.2.1 Fixed-point Binary Numbers

As the name implies, for fixed-point numbers, the location of the binary point remains in a fixed position for all arithmetic operations. Fixed-point binary numbers can be represented in several ways; the three most common methods are: (1) representation by *sign and magnitude*; (2) *one's complement* representation; and (3) *two's complement* representation.

For the sign and magnitude representation the binary word consists of the positive magnitude and a sign bit as the leading binary digit. For example, 0101*110 represents +5.75, and 1101*110 represents −5.75.

Positive numbers are represented by the sign and magnitude method but negative numbers are normally represented in either the one's complement or two's complement form. The one's complement of a binary number is simply formed by inverting (that is, changing 0s to 1s and vice versa) the sign and magnitude bits. For example, 0110*101 represents +6.625, and the corresponding one's complement representation of −6.625 is 1001*010. The two's complement of a binary number is determined by first forming the one's complement of the number, and then the two's complement is formed by adding 1 to the least significant bit of the one's complement representation. For example, the two's complement representation of −6.625 is 1001*011.

In arithmetic the multiplication of an integer by an integer produces an integer result. Alternatively the product of two fractions is a fraction and in this case arithmetic overflow is absent. In digital filter implementations it is necessary to approximate the $2w$-bit product of two w-bit numbers by a w-bit finite word length and—compared with integer arithmetic—this is more easily achieved using fractional arithmetic, the $2w$-bit product being rounded or truncated to w-bits. For truncation every bit after the wth bit is dropped. For rounding to w-bits, if the $(w + 1)$th bit is 0 we add 0 to the wth bit, but if the $(w + 1)$th bit is 1 we add 1 to the wth bit.

A signed decimal fraction may be represented by a $(w + 1)$-bit binary number in two's complement form, as follows

$$(N)_{10} = -C_0 + \sum_{j=1}^{w} C_j 2^{-j} \tag{4.2}$$

where C_0 (the sign-bit) and $C_j = 0$ or 1. Referring to equation 4.2, the largest positive number that can be represented is $1 - 2^{-w}$, and the largest negative number is -1.

The absolute error, E = (actual number $(N)_{10}$ − quantised number $(\bar{N})_{10}$), introduced into the binary number representation is less serious when rounding is used compared with when truncation is used, and consequently rounding is generally preferable in digital filter implementations. For rounding, the range of absolute error in the approximation is $-2^{-w/2} \leqslant E < 2^{-w/2}$, hence the longer the word length, the smaller the absolute error in the approximation.

When adding two fixed-point fractions it is possible that overflow will occur, thereby introducing an error into the sum. For example, $0*101 + 0*011 = 1*000$, and we see that overflow has occurred, therefore the answer is incorrect. Thus we see that there is a limitation on the range of numbers that can be represented using a fixed-point format. A much wider dynamic range of numbers can be realised using a floating-point format without the necessity of increasing word length.

4.2.2 Floating-point Binary Numbers

In digital systems binary floating-point numbers are normally represented in the form

$$(N)_{10} = 2^c m \tag{4.3}$$

where c is the *characteristic* (exponent) and m is the *mantissa* of the number. The characteristic and mantissa can be positive or signed. Unique representation of any number is achieved by restricting the range of the mantissa, that is

$$\tfrac{1}{2} \leqslant m < 1$$

and the corresponding floating-point binary number is said to be *normalised*.

In floating-point arithmetic multiplication and division are relatively simple, but in contrast addition and subtraction are more difficult to accomplish. Suppose that two floating-point numbers are

$$N_1 = 2^{c_1} m_1 \quad \text{and } N_2 = 2^{c_2} m_2$$

Then

$$N_1 \times N_2 = 2^{(c_1 + c_2)} (m_1 \times m_2)$$

and

$$\frac{N_1}{N_2} = 2^{(c_1 - c_2)} \left(\frac{m_1}{m_2}\right)$$

Note that if the operation of multiplication or division produces an answer such that $(\frac{1}{2} \leqslant m < 1)$ is not valid, then the characteristic of the result is changed to bring the mantissa into the range for normalised floating-point binary numbers. Example 4.1 illustrates this point.

Example 4.1
Suppose that $N_1 = 2.5$ and $N_2 = 1.25$; convert both to floating-point binary numbers, and then evaluate the product $N_1 \times N_2$; express the result as a normalised floating-point binary number.

SOLUTION
$$N_1 = 2.5 = 2^{10*0} 0*101$$
$$N_2 = 1.25 = 2^{1*0} 0*101$$
$$N_1 \times N_2 = 2^{(10*0)+(1*0)} [(0*101)(0*101)]$$
$$= 2^{11*0} (0*011001)$$

The mantissa is less than $\frac{1}{2}$, and in this case it can be brought into the desired range $(\frac{1}{2} \leqslant m < 1)$ by shifting the mantissa one place to the left and at the same time decrementing the characteristic, thus

$$N_1 \times N_2 = 2^{10*0} 0*11001 = 3.125$$

To add or subtract two floating-point binary numbers it is necessary to adjust the mantissa of the smaller number until the characteristics c_1 and c_2 are equal, and then the sum is formed by adding m_1 to m_2, or alternatively by subtracting m_2 from m_1 to obtain the difference. Example 4.2 illustrates the method.

Example 4.2
Using the values of N_1 and N_2 in example 4.1, evaluate the sum $N_1 + N_2$; express the result as a normalised floating-point binary number.

SOLUTION

N_2 is the smaller of the two numbers, and therefore $N_2 = 2^{1*0}.0*101$ must be modified to $N_2 = 2^{10*0}.0*0101$ so that $c_2 = c_1$, therefore

$$N_1 + N_2 = 2^{10*0}[(0*101) + (0*0101)]$$

$$= 2^{10*0}0*1111 = 3.75$$

In floating-point arithmetic relative error, ϵ, is more important than absolute error, E. Taking the floating-point number to have a mantissa rounded to k-bits with a relative error in the range $-2^{-k} \leqslant \epsilon < 2^{-k}$, then the floating-point approximation to $(N)_{10}$ is given by

$$(\bar{N})_{10} = (N)_{10}(1 + \epsilon) \quad \text{where } \epsilon = \frac{(\bar{N})_{10} - (N)_{10}}{(N)_{10}}$$

4.2.3 Comparison of Fixed-point and Floating-point Binary Numbers

When using a fixed-point representation of binary numbers truncation or rounding errors exist in arithmetic multiplication but not in arithmetic addition. In contrast, these errors are present for multiplication or addition of floating-point binary numbers.

In fixed-point format (equation 4.2) the dynamic range of the number is restricted to $-1 \leqslant (N)_{10} \leqslant (1 - 2^{-w})$, and arithmetic addition can produce overflow, thereby producing a result that exceeds the valid dynamic number range. In floating-point format the dynamic number range is relatively large. For example, the dynamic range of the 16-bit floating-point numbers illustrated in figure 4.1 is

$$2^{-65} \leqslant |(N)_{10}| < 2^{63}$$

and therefore with a very large dynamic range of numbers overflow is unlikely to occur in the filter implementation.

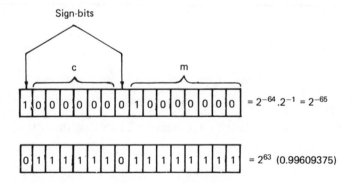

Figure 4.1 The dynamic range of a 16-bit floating-point number

For normalised fixed-point n-bit word lengths, or a floating-point k-bit mantissa, the ranges of errors are known, and it is generally assumed that within the appropriate range they have uniform probability density functions, as shown in figure 4.2.

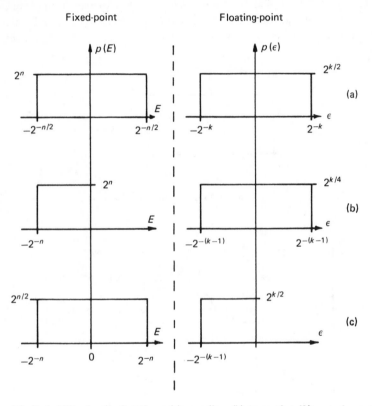

Figure 4.2   Probability density functions: (a) rounding; (b) truncation (2's complement); (c) truncation (1's complement or sign and magnitude)

4.3 QUANTISATION OF THE DIGITAL FILTER INPUT SIGNAL

In the implementation of a practical digital filter system a constituent component is the A/D converter (see figure 1.1), which normally produces fixed-point binary number representations of the input samples, $x^*(t)$, and this process is generally referred to as *quantisation* of the input signal. The difference between the actual input sample value and the fixed-point binary representation is commonly referred to as *quantisation noise*.

An A/D converter used with a microprocessor based digital filter system may typically have a $(w + 1)$-bit (includes sign bit) fixed-point output, with a corresponding resolution of about 1 part in 2^w. In this case the difference between adjacent quantisation levels is $Q = 1/2^w$ and for rounding the maximum possible quantisation error is $Q/2$. It is generally assumed that the probability density function for round-off quantisation error is uniform, as shown in figure 4.3.

Referring to figure 4.4 and assuming that the analogue signal is linear in the time interval t_x to t_y, then the corresponding quantisation error and squared quantisation error will be as shown in figure 4.5a and figure 4.5b respectively.

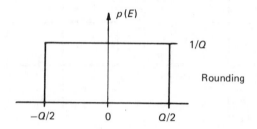

Figure 4.3 Probability density function for round-off quantisation error

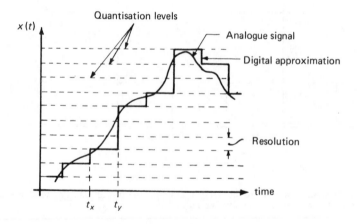

Figure 4.4 Quantisation of the filter input signal

The mean-square error (variance or average noise power) may be determined by finding an expression for the average value of the waveform shown in figure 4.6, that is

$$\text{variance} = \frac{2}{t_y} \int_0^{t_y/2} \left(\frac{Qt}{t_y}\right)^2 dt$$

Therefore variance

$$\sigma_i^2 = \frac{Q^2}{12} = \frac{2^{-2w}}{12} \tag{4.4}$$

It is worth noting that the mean-square noise error (equation 4.4) depends only on Q, and it is independent of the time interval $(t_y - t_x)$. Clearly the greater the number of bits in the A/D converter output, the smaller the value of Q, and the smaller the corresponding quantisation noise introduced by the conversion process.

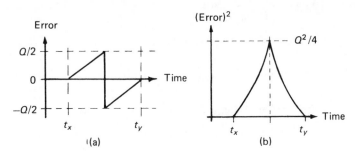

Figure 4.5   (a) The quantisation error; (b) the square quantisation error

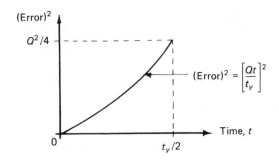

Figure 4.6   Waveform used to determine the mean-square error

The input to the digital filter is considered to be the sum of two component signals, namely, a noiseless input $x(n)T$ and a noise input $e(n)T$. That is, the quantised input signal is expressed as

$$x(n)T|_q = x(n)T + e(n)T \qquad (4.5)$$

In practice, if the amplitude of the input signal exceeds the dynamic number range of the A/D converter, then the waveform amplitude must be reduced to eliminate clipping. Thus in our model of the quantisation process we may introduce a scaling factor, F, such that

$$x(n)T|_q = Fx(n)T + e(n)T \qquad (4.6)$$

where $0 < F < 1$. In this case, and assuming that the quantisation errors are produced by rounding, the signal-to-noise ratio (SNR) of $x(n)T|_q$ is

$$SNR = 10 \log_{10} \left(\frac{F^2 \sigma_x{}^2}{\sigma_i{}^2} \right) dB \qquad (4.7)$$

where $F^2 \sigma_x^2$ is the variance of $Fx(n)T$, and σ_i^2 is the variance of the quantisation noise (see equation 4.4). As a rule-of-thumb, for negligible clipping, F is generally set at $1/(5\sigma_x)$, thus for this case

$$\text{SNR} = 10 \log_{10} \left[\frac{1}{(25\sigma_i^2)} \right] \text{dB}$$

$$= (-10 \log_{10} 25 - 10 \log_{10} \sigma_i^2) \, \text{dB}$$

$$= \left[-13.9794 - 10 \log_{10} \left(\frac{2^{-2w}}{12} \right) \right] \text{dB}$$

$$= (-13.9794 - 10 \log_{10} 2^{-2w} + 10 \log_{10} 12) \, \text{dB}$$

$$= (-13.9794 + 20w \log_{10} 2 + 10.7918$$

Therefore

$$\text{SNR} = (6.02w - 3.1876) \, \text{dB} \tag{4.8}$$

Using equation 4.8 we see that an 8-bit A/D converter would have a SNR ≈ 45 dB.

The variance at the filter output, σ_0^2 due to input quantisation (defined by equation 4.5) may be obtained using Parseval's theorem for discrete systems,[7] that is

$$\sigma_0^2 = \frac{Q^2}{12} \sum_{i=0}^{\infty} [g(i)T]^2 = \frac{Q^2}{12} \frac{1}{j2\pi} \oint G(Z)G^*(Z) \frac{dZ}{Z} \tag{4.9}$$

Example 4.3 illustrates the method.

Example 4.3
A first-order recursive digital filter has a pulse transfer function, $G(Z) = Z/(Z - A)$ where $|A| < 1$. Using Parseval's theorem (equation 4.9), derive an expression for the variance at the filter's output due to A/D converter input quantisation. Comment on the result.

SOLUTION
The impulse response of the filter, $g(i)T$, is equal to $\mathbf{Z}^{-1}[G(Z)]$, which may be obtained by referring to the table of Z-transforms (table 1.2), thus $g(i)T = A^i$ and therefore $[g(i)T]^2 = A^{2i}$. Using equation 4.9 we obtain

$$\sigma_0^2 = \frac{Q^2}{12} \sum_{i=0}^{\infty} A^{2i}$$

$$= \frac{Q^2}{12} (1 + A^2 + A^4 + A^6 + \ldots)$$

Therefore

$$\sigma_0{}^2 = \frac{Q^2}{12} \frac{1}{(1 - A^2)}$$

Clearly as $A^2 \to 1$ then $\sigma_0{}^2 \to \infty$, and as $A^2 \to 0$ then $\sigma_0{}^2 \to Q^2/12$. Thus we see that in this example the location of the pole has some significant effect on the value of $\sigma_0{}^2$.

Round-off error analysis in digital filter implementations can be developed by considering the various realisation structures used to represent the digital filter pulse transfer function. These are described in the following section.

4.4 REALISATION STRUCTURE CONSIDERATIONS[8,9,10,23]

4.4.1 Quantisation Effects due to Rounded Fixed-point Arithmetic

The pulse transfer function, $G(Z)$, of a recursive digital filter may be expressed as

$$G(Z) = \frac{Y(Z)}{X(Z)} = \frac{\displaystyle\sum_{i=0}^{K} a_i Z^{-i}}{1 + \displaystyle\sum_{i=1}^{J} b_i Z^{-i}} \qquad (4.10)$$

The linear difference equation corresponding to equation 4.10 is

$$y(n)T = \sum_{i=0}^{K} a_i x(n-i)T - \sum_{i=1}^{J} b_i y(n-i)T \qquad (4.11)$$

The realisation structure shown in figure 4.7 represents a *direct form* implementation of equation 4.11, and it is seen that a kth order filter requires $2k$ delay operations. It is possible to manipulate $G(Z)$ to produce a *canonic direct form* realisation such that the filter implementation only requires k delay operations— this is achieved as follows. Let $P(Z) = G(Z)Q(Z)$, and therefore

$$P(Z) = \frac{Y(Z)Q(Z)}{X(Z)} \qquad (4.12)$$

Now denoting the ratio $Q(Z)/X(Z)$ as $R(Z)$, then equation 4.12 may be written as

$$P(Z) = Y(Z)R(Z) \qquad (4.13)$$

The linear difference equation corresponding to $R(Z)X(Z) = Q(Z)$ is

$$r(n)T \left(1 + \sum_{i=1}^{J} b_i Z^{-i}\right) = q(n)T$$

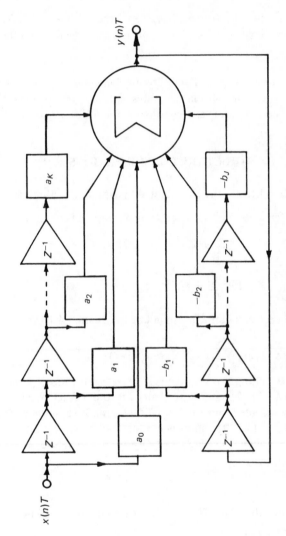

Figure 4.7 Realisation structure for a *direct form* implementation of a recursive digital filter

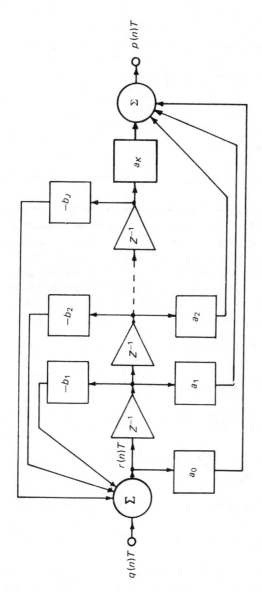

Figure 4.8 The *canonic direct form* realisation structure

that is

$$r(n)T = q(n)T - \sum_{i=1}^{J} b_i r(n - i)T \qquad (4.14)$$

and the linear difference equation corresponding to equation 4.13 is

$$p(n)T = y(n)Tr(n)T = \left(\sum_{i=0}^{K} a_i Z^{-i} \right) r(n)T$$

that is

$$p(n)T = \sum_{i=0}^{K} a_i r(n - i)T \qquad (4.15)$$

Using equations 4.14 and 4.15 the canonic direct form realisation structure may be drawn, as shown in figure 4.8.

Now at this point it is appropriate to note that the direct form of realisation structure has a total of $(J + K + 1)$ multiplication operations involved in the computation of $y(n)T$—see equation 4.11 and figure 4.7. Thus, if we assume that we are dealing with fixed-point $2w$-bit products, which are rounded to a w-bit result, then clearly quantisation errors will exist. These errors, as we have already seen in section 4.3, may be regarded as random white noise with zero mean and with a variance equal to $Q^2/12$. Since the $(J + K + 1)$ quantisation errors are independent and white noise, their sum $\epsilon(n)T$, is

$$\epsilon(n)T = \frac{(J + K + 1) Q^2}{12} \qquad (4.16)$$

and the computed (rounded) filter output, $y_r(n)T$ is

$$y_r(n)T = \sum_{i=0}^{K} a_i x(n - i)T - \sum_{i=1}^{J} b_i y_r(n - i)T + \epsilon(n)T \qquad (4.17)$$

The output error in the filter output may be defined as

$$E(n)T = y_r(n)T - y(n)T \qquad (4.18)$$

Now substituting equations 4.11 and 4.17 in equation 4.18 we obtain

$$\epsilon(n)T = E(n)T + \sum_{i=1}^{J} b_i E(n - i)T \qquad (4.19)$$

This linear difference equation describes a linear discrete system with an input $\epsilon(n)T$ and a corresponding output $E(n)T$, that is, a system having a pulse transfer function

$$\frac{E(Z)}{\epsilon(Z)} = \frac{1}{1 + \sum_{i=1}^{J} b_i Z^{-i}} = H(Z) \qquad (4.20)$$

Referring back to equation 4.10 we see that $1 + \sum_{i=1}^{J} b_i Z^{-1}$ is the denominator of the pulse transfer function, $G(Z)$. Thus we see that $H(Z) = 1/B(Z)$, where $B(Z)$ is the denominator of $G(Z)$. Consequently we may deduce that $B(Z)$ is the part of the original pulse transfer function, $G(Z)$, that contributes to the noise in the filter's output signal (see figure 4.9). Using Parseval's theorem (equation 4.9) the average output noise power (variance) may be determined, and is given by

$$\sigma_T^2 = \frac{(J + K + 1)Q^2}{12} \frac{1}{j2\pi} \oint \frac{1}{B(Z)} \frac{1}{B^*(Z)} \frac{dZ}{Z} \tag{4.21}$$

Example 4.4 illustrates the method.

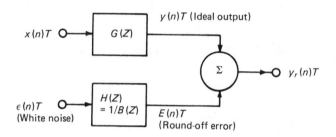

Figure 4.9 Block diagram of digital filter which takes account of round-off error

Example 4.4
A digital filter has a pulse transfer function

$$G(Z) = \frac{0.75}{(Z - 0.4)(Z - 0.5)}$$

Determine an expression for the average output noise power in a *direct form* implementation.

SOLUTION

$$B(Z) = (Z - 0.4)(Z - 0.5)$$

$$B^*(Z) = (Z^{-1} - 0.4)(Z^{-1} - 0.5)$$

$$= \frac{Z^2}{(1 - 0.4Z)(1 - 0.5Z)}$$

Order of numerator of $G(Z)$, $K = 0$. Order of denominator of $G(Z)$, $J = 2$. Using equation 4.21 we obtain

$$\sigma_T^2 = \frac{(2 + 0 + 1)Q^2}{12} \frac{1}{j2\pi} \oint \frac{Z^2 \, dZ}{(Z - 0.4)(Z - 0.5)(1 - 0.4Z)(1 - 0.5Z)Z}$$

$$= \frac{Q^2}{4} \frac{1}{j2\pi} \oint \frac{Z}{(Z - 0.4)(Z - 0.5)(1 - 0.4Z)(1 - 0.5Z)} \, dZ$$

The integration may be evaluated by summing the residues due to the singularities (poles) within the contour of integration (unit-circle in Z-plane). That is

$$\sigma_T^2 = \frac{Q^2}{4} \frac{1}{j2\pi} \times j2\pi \, \Sigma \, (\text{residues})$$

The poles of $B(Z)B^*(Z)$ that lie inside the unit-circle are located at $Z = 0.4$ and $Z = 0.5$, and their respective residues are

$$\text{residue}_1 = \frac{0.4}{(0.4 - 0.5)(1 - 0.4^2)(1 - 0.5 \, 0.4)} = -5.9524$$

$$\text{residue}_2 = \frac{0.5}{(0.5 - 0.4)(1 - 0.4 \, 0.5)(1 - 0.5^2)} = 8.3333$$

Therefore

$$\sigma_T^2 = \frac{Q^2}{4} (8.3333 - 5.9524)$$

that is

$$\sigma_T^2 = 0.5952 \, Q^2 \tag{4.22}$$

Up to this point we have considered the case when the filter is recursive. However, for a non-recursive filter the b_i coefficients are zero, and consequently $J = 0$ and $B(Z) = 1$ (see equation 4.10), and therefore the integral in equation 4.21 may be evaluated, and it will be equal to $j2\pi$. Thus the average output noise power of a non-recursive digital filter is

$$\sigma_T^2 = \frac{(K + 1)Q^2}{12} \tag{4.23}$$

To realise a digital filter in a *cascade form* (see figure 4.10), the pulse transfer function, $G(Z)$, is factorised to the form

$$G(Z) = \prod_{i=1}^{R} \left(\frac{\alpha_{0i} + \alpha_{1i} Z^{-1} + \alpha_{2i} Z^{-2}}{1 + \beta_{1i} Z^{-1} + \beta_{2i} Z^{-2}} \right) \tag{4.24}$$

that is

$$G(Z) = \prod_{i=1}^{R} F_i(Z) \tag{4.25}$$

where

$$F_i(Z) = \left(\frac{\alpha_{0i} + \alpha_{1i} Z^{-1} + \alpha_{2i} Z^{-2}}{1 + \beta_{1i} Z^{-1} + \beta_{2i} Z^{-2}} \right)$$

$i = 1, 2, 3, \ldots, R$. To include the effect of round-off noise the ideal cascade form of realisation shown in figure 4.10 may be modified as shown in figure 4.11, where $\beta_i(Z)$ is the denominator of $F_i(Z)$. The noise inputs, $\epsilon_i(n)T$, $i = 1, 2, 3, \ldots, R$, will each produce a quantisation error component, σ_i^2, in the output signal, and the total quantisation error, σ_T^2, is

$$\sigma_T^2 = \sum_{i=1}^{R} \sigma_i^2 \tag{4.26}$$

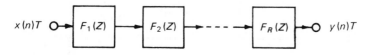

$$x(n)T \; \text{O} \rightarrow \boxed{F_1(Z)} \longrightarrow \boxed{F_2(Z)} \rightarrow - - - \rightarrow \boxed{F_R(Z)} \rightarrow \text{O} \; y(n)T$$

Figure 4.10 Block diagram of *cascaded* sections to form composite digital filter

By inspection of figure 4.11 we see that the pulse transfer function seen by each noise input $\epsilon_i(n)T$ is

$$\left. \begin{aligned} &\frac{1}{\beta_i(Z)} \left[\prod_{k=i+1}^{R} F_k(Z) \right] \quad \text{for } i = 1, 2, 3, \ldots, (R-1) \\ &\frac{1}{\beta_R(Z)} \quad \text{for } i = R \end{aligned} \right\} \tag{4.27}$$

It follows that the ith quantisation error component has a variance value given by Parseval's theorem, that is

$$\left. \begin{aligned} \sigma_i^2 &= \frac{Q^2}{12} \frac{1}{j2\pi} \oint \frac{1}{\beta_i(Z)} \frac{1}{\beta_i{}^*(Z)} \left[\prod_{k=i+1}^{R} F_k(Z) F_k{}^*(Z) \right] \frac{dZ}{Z} \\ &\text{for } i = 1, 2, 3, \ldots, (R-1), \text{ or} \\ \sigma_i^2 &= \frac{Q^2}{12} \frac{1}{j2\pi} \oint \frac{1}{\beta_R(Z)} \frac{1}{\beta_R{}^*(Z)} \frac{dZ}{Z} \end{aligned} \right\} \tag{4.28}$$

for $i = R$. The following worked example illustrates the method.

Example 4.5
For the pulse transfer function given in example 4.4, determine the average output noise power in a *cascade form* implementation.

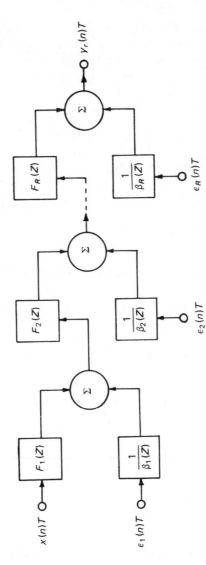

Figure 4.11 Block diagram of a *cascade realisation* of a digital filter which takes account of round-off error

SOLUTION

$$G(Z) = \frac{0.75}{(Z - 0.4)(Z - 0.5)}$$

(see example 4.4). In cascade form, $G(Z)$ may be considered to be a pulse transfer function $F_1(Z)$ followed by a second pulse transfer function $\beta_2(Z)$, where

$$F_1(Z) = \frac{0.75}{(Z - 0.4)} \quad \text{and } \beta_2(Z) = \frac{1}{(Z - 0.5)}$$

that is

$$G(Z) = F_1(Z)\beta_2(Z)$$

Using equation 4.28 we obtain

$$\sigma_1{}^2 = \frac{Q^2}{12} \frac{1}{j2\pi} \oint \frac{1}{(Z - 0.4)} \frac{1}{(Z^{-1} - 0.4)} \frac{1}{(Z - 0.5)} \frac{1}{(Z^{-1} - 0.5)} \frac{dZ}{Z}$$

Therefore

$$\sigma_1{}^2 = \frac{Q^2}{12} \frac{1}{j2\pi} \oint \frac{5Z}{(Z - 0.4)(Z - 0.5)(2.5 - Z)(2 - Z)} dZ$$

There are two poles within the contour of integration, namely at $Z = 0.4$ and at $Z = 0.5$, and their corresponding residues are

$$\text{residue}_1 = \frac{5 \times 0.4}{(0.4 - 0.5)(2.5 - 0.4)(2 - 0.4)} = -5.9524$$

$$\text{residue}_2 = \frac{5 \times 0.5}{(0.5 - 0.4)(2.5 - 0.5)(2 - 0.5)} = 8.3333$$

Therefore we obtain

$$\sigma_1{}^2 = \frac{Q^2}{12}(-5.9524 + 8.3333)$$

that is

$$\sigma_1{}^2 = 0.1984 Q^2$$

Again using equation 4.28 we obtain

$$\sigma_2{}^2 = \frac{Q^2}{12} \frac{1}{j2\pi} \oint \frac{1}{(Z - 0.5)} \frac{1}{(Z^{-1} - 0.5)} \frac{dZ}{Z}$$

Therefore

$$\sigma_2{}^2 = \frac{Q^2}{12} \frac{1}{j2\pi} \oint \frac{2}{(Z - 0.5)(2 - Z)} dZ$$

There is a single pole within the contour of integration at $Z = 0.5$, and its corresponding residue is

$$\text{residue} = \frac{2}{(2 - 0.5)} = 1.3333$$

Therefore

$$\sigma_2{}^2 = \frac{Q^2}{12} \times 1.3333 = 0.111 Q^2$$

Now using equation 4.26 we obtain

$$\sigma_T{}^2 = \sigma_1{}^2 + \sigma_2{}^2 = 0.3095 Q^2 \tag{4.29}$$

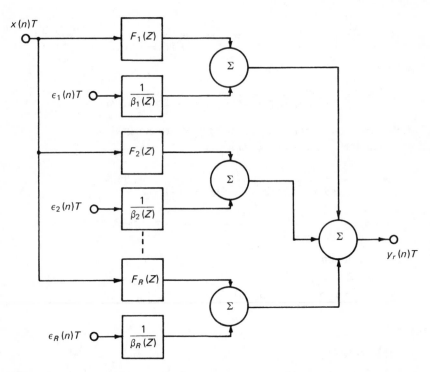

Figure 4.12 Block diagram of a *parallel realisation* of a digital filter which takes account of round-off error

To realise a digital filter in *parallel form* (see figure 4.12) the pulse transfer function, $G(Z)$, is written as a partial fraction expansion, thus

$$G(Z) = \sum_{i=1}^{R} F_i(Z)$$

where

$$F_i(Z) = \left(\frac{\alpha_{0i} + \alpha_{1i} Z^{-1} + \alpha_{2i} Z^{-2}}{1 + \beta_{1i} Z^{-1} + \beta_{2i} Z^{-2}} \right)$$

$i = 1, 2, 3, \ldots, R$. The noise inputs, $\epsilon_i(n)T$, $i = 1, 2, 3, \ldots, R$, each produce a quantisation error component, σ_i^2, in the output signal. The total quantisation error at the filter output is the sum of the σ_i^2s, that is

$$\sigma_T^2 = \sum_{i=1}^{R} \sigma_i^2 \qquad (4.30)$$

where

$$\sigma_i^2 = \frac{Q^2}{12} \frac{1}{j2\pi} \oint \frac{1}{\beta_i(Z)} \frac{1}{\beta_i^*(Z)} \frac{dZ}{Z} \qquad (4.31)$$

for $i = 1, 2, \ldots, R$. The following worked example illustrates the method.

Example 4.6
For the pulse transfer function given in example 4.4, determine the average output noise power in a *parallel form* implementation.

SOLUTION

$$G(Z) = \frac{0.75}{(Z - 0.4)(Z - 0.5)}$$

(see example 4.4). Now expanding into partial fractions we obtain

$$G(Z) = \frac{7.5}{Z - 0.5} - \frac{7.5}{Z - 0.4}$$

Using equation 4.31 we obtain

$$\sigma_1^2 = \frac{Q^2}{12} \frac{1}{j2\pi} \oint \frac{1}{(Z - 0.4)} \frac{1}{(Z^{-1} - 0.4)} \frac{dZ}{Z}$$

Therefore

$$\sigma_1^2 = \frac{Q^2}{12} \frac{1}{j2\pi} \oint \frac{2.5}{(Z - 0.4)(2.5 - Z)} \, dZ$$

One pole is within the contour of integration at $Z = 0.4$, and its corresponding residue is

$$\text{residue} = \frac{2.5}{2.5 - 0.4} = 1.1905$$

Therefore

$$\sigma_1{}^2 = \frac{Q^2}{12} \times 1.1905 = 0.0992Q^2$$

Similarly,

$$\sigma_2{}^2 = \frac{Q^2}{12} \frac{1}{j2\pi} \oint \frac{1}{(Z - 0.5)} \frac{1}{(Z^{-1} - 0.5)} \frac{dZ}{Z}$$

Therefore

$$\sigma_2{}^2 = \frac{Q^2}{12} \frac{1}{j2\pi} \oint \frac{2}{(Z - 0.5)(2 - Z)} dZ$$

One pole is within the contour of integration, at $Z = 0.5$, and its corresponding residue is

$$\text{residue} = \frac{2}{2 - 0.5} = 1.3333$$

Therefore

$$\sigma_2{}^2 = \frac{Q^2}{12} \times 1.3333 = 0.1111Q^2$$

From equation 4.30 we obtain

$$\sigma_T{}^2 = \sigma_1{}^2 + \sigma_2{}^2 = 0.2103Q^2$$

It is worth noting that in comparing the answers to examples 4.4, 4.5 and 4.6 for this particular pulse transfer function, the *parallel form* of implementation gave the least quantisation (round-off) error.

4.4.2 Effect of Finite Word Length on Stability and Frequency Response Characteristics

It was briefly mentioned in section 1.6.2 that if one or more poles of $G(Z)$ are located close to the circumference of the unit-circle in the Z-plane, thereby rendering $G(Z)$ marginally stable, it will be necessary to investigate how a small change in one of the denominator coefficients, b_k, $1 \leqslant k \leqslant m$ (see equation 1.35) may result in one or more poles moving outside the unit-circle, thereby making the filter unstable. This small change is brought about by inaccurate representation of the coefficients using finite, fixed-point, processor word lengths.

The exact pole positions of the ideal filter may be computed from its characteristic equation, namely from

$$1 + \sum_{k=1}^{m} b_k Z^{-k} = 0 \tag{4.32}$$

If we now assume that one of the filter coefficients, b_i, is changed to a new value equal to $b_i + \Delta b_i$, the characteristic equation of the filter in this case is

$$1 + \sum_{k=1}^{m} b_k Z^{-k} + \Delta b_i Z^{-i} = 0 \qquad (4.33)$$

Also if we assume that the sampling frequency is relatively high, such that the poles are within the unit-circle close to $Z = 1$, then we seek the value of Δb_i which will result in a pole moving outside the unit-circle, and since this occurs at $Z = 1$, equation 4.33 becomes

$$\Delta b_i = 1 + \sum_{k=1}^{m} b_k \qquad (4.34)$$

By comparing Δb_i with the largest b_k coefficient the required resolution may be determined,[11] and hence the required processor word length will be known. The following worked example illustrates the method.

Example 4.7
A digital filter designed by the impulse-invariant design method (see chapter 2), which is based on a prototype second-order Butterworth lowpass filter with cutoff at 1 rad/s and sampling frequency equal to 30.2 rad/s, has a pulse transfer function, $G(Z)$, equal to $(0.0373Z)/(Z^2 - 1.7Z + 0.745)$. Calculate the minimum word length for stability to be maintained, assuming that filter coefficients are rounded.

SOLUTION
Using equation 4.34 we obtain

$$\Delta b_i = 1 - 1.7 + 0.745 = 0.045$$

and the corresponding quantisation interval $= 2.(0.045) = 0.09$. Now comparing this with the largest b_k coefficient we may write

$$\frac{1.7}{0.09} \leqslant 2^{(w-1)}$$

Thus the minimum word length to maintain stability is $w = 6$. However, note that if coefficients are truncated rather than rounded the word length is increased by 1 to $w = 7$.

An alternative method of calculating the processor word length to maintain stability has been reported by Kuo and Kaiser.[11] Assuming truncation rather than rounding they showed that for an Nth order filter having distinct poles at $(\cos \omega_k T - j \sin \omega_k T)$, $k = 1, 2, \ldots, N$, the number of processor bits must be

$$w = \text{smallest integer exceeding} \left\{ -\log_2 \left[\frac{5\sqrt{N}}{2^{N+2}} \prod_{k=1}^{N} (\omega_k T) \right] \right\} \quad (4.35)$$

At this point it is appropriate to verify the statement made in the last sentence of the solution to example 4.7. That is, using equation 4.35 we will show that to maintain stability for the digital filter pulse transfer function quoted in example 4.7, the processor word length is $w = 7$ (see example 4.8.)

Example 4.8
Repeat example 4.7 assuming that filter coefficients are truncated rather than rounded.

SOLUTION
The Z-plane poles of the digital filter are nominally at $Z = (0.85 \pm j0.15)$. In this case the filter's pulse transfer function has a second-order dominator, therefore $N = 2$. Since the two poles are a complex conjugate pair, then

$$\omega_1 T = \omega_2 T = 0.1747 \text{ rad} = \left(\tan^{-1} \frac{0.15}{0.85} \right)^{\circ}$$

Substituting N, $\omega_1 T$ and $\omega_2 T$ in equation 4.35 we obtain

$$w = \text{smallest integer exceeding} [-\log_2 (0.0135)]$$

Let

$$-\log_2 (0.0135) = x$$

then

$$\log_2 (0.0135) = -x$$

Therefore

$$0.0135 = 2^{-x}$$

that is

$$\frac{1}{0.0135} = 2^x = 74.074$$

Since $x \approx 6.21$ and since w must be an integer it follows that

$$w = 7$$

We see that the calculations in example 4.7 or example 4.8 yield the word length that ensures that stability is maintained. However, there is no guarantee that with this value of w the filter will meet its specified frequency response characteristic. Generally, a rule of thumb is used whereby three or four bits are added to the minimum number calculated to ensure stability, and normally this word length provides sufficient accuracy to meet the frequency response specification.

The movement of the filter's poles and zeros due to coefficient rounding may be calculated, and the corresponding changes in the frequency response characteristic can then be determined. Taking the poles of $G(Z)$ to be positioned on the Z-plane at p_l, $l = 1, 2, \ldots, J$ (see equation 4.10). Then it has been shown by Kaiser[12] that the poles of the quantised filter move to $p_l + \Delta p_l$, where

$$\Delta p_l = \sum_{i=1}^{J} \frac{p_l^{(i+1)}}{\prod_{\substack{n=1 \\ n \neq i}}^{J} (1 - p_l/p_n)} \Delta b_i \qquad (4.36)$$

and where Δb_i is the change in the value of the b_i filter coefficient. Similar results can be obtained for the movement in the filter zeros. For high order filters a significant saving in the number of coefficient bits can be achieved by using the cascade or parallel form of realisation. The direct form of realisation should only be considered for first or second-order filters. For higher order filters it is advisable to make the saving offered by the cascade or parallel forms.

The following example illustrates that different sensitivities are produced by different forms of realisation. Consider a second-order digital filter having the pole-zero representation shown in figure 4.13. The corresponding pulse transfer function of the filter is

$$G(Z) = \frac{Z^2}{(Z - p_1)(Z - p_2)} \qquad (4.37)$$

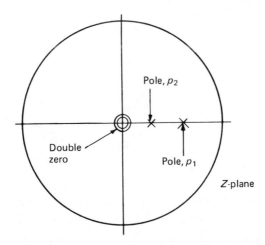

Figure 4.13 Z-plane pole–zero diagram of the second order digital filter defined by equation 4.37

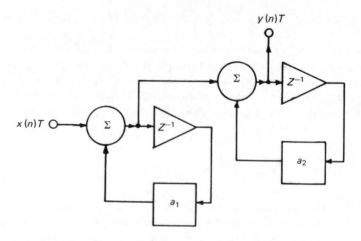

Figure 4.14   Cascade realisation for the digital filter defined by equation 4.37

Therefore

$$G(Z) = \frac{Z^2}{Z^2 - Z(p_1 + p_2) + p_1 p_2} \tag{4.38}$$

The pulse transfer function given by equation 4.37 can be implemented in cascade form as shown in figure 4.14, where the filter coefficients are

$$a_1 = p_1 \tag{4.39}$$

$$a_2 = p_2 \tag{4.40}$$

and it follows that

$$\frac{\partial p_1}{\partial a_1} = 1 \tag{4.41}$$

$$\frac{\partial p_1}{\partial a_2} = 0 \tag{4.42}$$

$$\frac{\partial p_2}{\partial a_1} = 0 \tag{4.43}$$

$$\frac{\partial p_2}{\partial a_2} = 1 \tag{4.44}$$

Alternatively, the pulse transfer function given by equation 4.38 can be implemented in *direct form* as shown in figure 4.15, where the filter coefficients are

$$b_1 = p_1 + p_2 \tag{4.45}$$

$$b_2 = p_1 \cdot p_2 \tag{4.46}$$

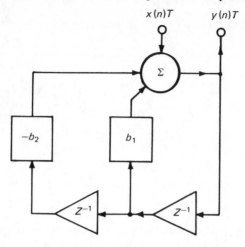

Figure 4.15   Direct-form realisation for the digital filter defined by equation 4.37

and since b_1 and b_2 are independent variables it follows that

$$\frac{\partial p_1}{\partial b_2} + \frac{\partial p_2}{\partial b_2} = 0 \tag{4.47}$$

$$\frac{\partial p_1}{\partial b_1} + \frac{\partial p_2}{\partial b_1} = 1 \tag{4.48}$$

$$p_1 \frac{\partial p_2}{\partial b_2} + p_2 \frac{\partial p_1}{\partial b_2} = 1 \tag{4.49}$$

$$p_1 \frac{\partial p_2}{\partial b_1} + p_2 \frac{\partial p_1}{\partial b_1} = 0 \tag{4.50}$$

Now manipulating equations 4.47 to 4.50 we obtain

$$\frac{\partial p_1}{\partial b_1} = \frac{p_1}{p_1 - p_2} \tag{4.51}$$

$$\frac{\partial p_1}{\partial b_2} = \frac{1}{p_2 - p_1} \tag{4.52}$$

$$\frac{\partial p_2}{\partial b_1} = \frac{p_1}{p_2 - p_1} \tag{4.53}$$

$$\frac{\partial p_2}{\partial b_2} = \frac{1}{p_1 - p_2} \tag{4.54}$$

Thus for this filter we see that

$$\frac{\partial p_1}{\partial b_1} > \frac{\partial p_1}{\partial a_1}$$

$$\frac{\partial p_1}{\partial b_2} > \frac{\partial p_1}{\partial a_2}$$

$$\frac{\partial p_2}{\partial b_1} > \frac{\partial p_2}{\partial a_1}$$

$$\frac{\partial p_2}{\partial b_2} > \frac{\partial p_2}{\partial a_2}$$

Hence we may deduce that, compared with the pole movements produced by changes in a_1 and a_2, the changes in b_1 and b_2 produce larger variation in the pole movements. This means that the direct form of realisation, when compared with the cascade form, is more sensitive to coefficient quantisation.

A method of determining the minimum word length required to hold a given tolerance on a pole position is by finding the *sensitivity coefficient*,[11] γ, which is defined as

$$\gamma = \frac{\% \text{ change in position of pole}, p}{\% \text{ change in coefficient}, b_k} \qquad (4.55)$$

This method is illustrated in the following worked example.

Example 4.9
For the pulse transfer function, $G(Z)$, given in example 4.7, and using the sensitivity to changes in b_2, determine the minimum word length required to hold the Z-plane poles to within 0.5 per cent of their nominal values.

SOLUTION
For the given pulse transfer function of example 4.7 the Z-plane poles are nominally at

$$Z = (0.85 \pm j0.15)$$

that is

$$p_1 = (0.85 + j0.15) \quad \text{and} \quad p_2 = (0.85 - j0.15)$$

and therefore from equation 4.54.

$$\gamma = \frac{\partial p_2}{\partial b_2} = \frac{1}{p_1 - p_2} = -j3.33$$

Thus to hold the pole position to 0.5 per cent, coefficient b_2 must be held to

$$\frac{0.5}{3.33}\ \%$$

(see equation 4.55) that is, b_2 must be held to 0.15 per cent. Hence the required resolution is 2(0.15 per cent) = 0.3 per cent. However, for a fixed-point binary fraction represented by w-bits after the binary point, the resolution is $(2^{-w} \times 100)$ per cent. For $w = 8$ the resolution is 0.39 per cent, and for $w = 9$ the resolution is 0.195 per cent. Clearly the required resolution of 0.3 per cent falls between these two values, but since w is integer we must choose $w = 9$ to satisfy the given specification.

The digital filter in example 4.7 was designed using the impulse-invariant design method (see section 2.2.2) that is, the prototype second-order Butterworth low-pass filter

$$G(S) = \frac{1}{b}\left[\frac{b}{(S+a)^2+b^2}\right]$$

was transformed via this method to

$$G(Z) = \frac{[(T/b)\,e^{-aT}\sin bT]\,Z}{Z^2 - [2\,e^{-aT}\cos bT]\,Z + e^{-2aT}} \tag{4.56}$$

If the sampling frequency of the filter in example 4.7 is increased to 50 rad/s, the corresponding pulse transfer function of the digital filter is obtained via equation 4.56, hence in this case

$$G(Z) = \frac{0.0145Z}{Z^2 - 1.815Z + 0.837}$$

To hold the Z-plane poles to within 0.5 per cent of their nominal values a word length $w = 9$ is required (see example 4.9 for method of calculating w).

In general, higher sampling rates require a corresponding increase in the processor word length, and in the practical implementation of the filter a compromise between word length requirements and the permitted error in the approximation of $G(S)$ by $G(Z)$ will normally be necessary.

4.5 LIMIT CYCLE OSCILLATIONS[13,14,15,16]

When the input to a digital filter is constant, or zero, the fixed-point finite word length arithmetic rounding errors cannot be assumed to be uncorrelated random variables (independent additive white noise components). In this case the *round-off noise* is significantly dependent on the input signal, and its effect is easily seen by studying a simple example for a first-order recursive filter. Consider the following difference equation

$$y(n)T = Ky(n - 1)T + x(n)T \qquad\qquad (4.57)$$

Now assuming that the input to the filter is turned off, it follows that $x(n)T = 0$. Suppose that $K = 0.93$, then equation 4.57 becomes

$$y(n)T = 0.93\, y(n - 1)T \qquad\qquad (4.58)$$

and if the filter's initial condition is $y(-1)T = 11$, then it is possible to calculate the exact values of $y(n)T$ (obtained from equation 4.58 using infinite precision arithmetic) and the rounded values of $y(n)T$ [$y(n)T$ rounded to the nearest integer]. Table 4.1 shows a comparison of the exact and rounded $y(n)T$ values, and it is seen that the exact value of $y(n)T$ decays exponentially to zero, but the rounded value of $y(n)T$ reaches a steady-state value of 7. This is an example of a digital filter limit cycle output response to zero input. It is interesting to observe the filter's output response to zero input when the initial condition is $y(-1)T = 8$; table 4.2 shows the corresponding result. Referring to table 4.2 it is seen that the rounded value of the filter output is a finite constant for all values of n in the range $0 \leqslant n \leqslant \infty$. In fact, any initial condition for $y(-1)T$ existing in the range $-D \leqslant y(-1)T \leqslant D$, where D is the largest integer satisfying $D \leqslant \frac{1}{2}(1 - |K|)^{-1}$, will produce this effect, which is often referred to as the *deadband effect*.[17] In the above example the deadband effect will be present in the range $-7 \leqslant y(-1)T \leqslant 7$.

**Table 4.1 Limit cycle output response to zero input for
the digital filter defined by equation 4.58;
initial condition $y(-1)T = 11$**

n	*Exact $y(n)T$*	*Rounded $y(n)T$*
0	10.23	10
1	9.5139	9
2	8.847927	8
3	8.22857211	7
4	7.6525720623	7
5	7.116892017939	7

Let us now consider the filter defined by equation 4.57 with $K = -0.93$, $y(-1)T = 0$ and $x(n)T = \{12, 0, 0, 0, \ldots\}$; the filter's corresponding impulse response is summarised in table 4.3. Referring to table 4.3 we see that the filter oscillates with frequency equal to $\omega_s/2$ and amplitude ± 7.

In some practical applications the effects presented in tables 4.1, 4.2 and 4.3 are undesirable. For example, if a telephone network has a zero-level input signal, then it is possible that the finite output signal, produced by the deadband effect, will be unacceptable to the subscriber. The deadband effect can sometimes be reduced to an acceptable degree by increasing the word length of the processor. However, if this is not practicable, then the deadband effect can be reduced by adding a small *dither signal*[18] of the form $(-1)^n d$, where $d \approx \frac{1}{2}$ times the quantisa-

Table 4.2 Limit cycle output response to zero input for
the digital filter defined by equation 4.58;
initial condition $y(-1)T = 8$

n	$y(n)T$ Before Rounding	$y(n)T$ After Rounding to the Nearest Integer
0	7.44	7
1	6.51	7
2	6.51	7
.	.	.
.	.	.
.	.	.
∞	6.51	7

Table 4.3 Limit cycle output response to impulse input
$\{12, 0, 0, \ldots\}$ for the digital filter defined by
equation 4.57; initial condition $y(-1)T = 0$

n	$y(n)T$ Before Rounding	$y(n)T$ After Rounding to the Nearest Integer
0	12	12
1	−11.16	−11
2	10.23	10
3	− 9.3	− 9
4	8.37	8
5	− 7.44	− 7
6	6.51	7
7	− 6.51	− 7
8	6.51	7
9	− 6.51	− 7

tion interval, to the $y(n)T$ value before rounding is carried out. The effect of
adding the dither signal is to assist the filter output in jumping across the
quantisation threshold, thereby making the filter achieve a small steady-state
oscillation about the desired final output value. A dither signal may be added to
the right-hand side of equation 4.58, thus giving a difference equation of the form

$$y(n)T = 0.93\, y(n - 1)T + (-1)^n d$$

and choosing $d = 0.5$ we may write

Table 4.4 Limit cycle response to zero input for the
digital filter with dither signal (equation 4.59);
initial condition $y(-1)T = 11$

n	$y(n)T$ Before Rounding	$y(n)T$ After Rounding to the Nearest Integer
0	10.73	11
1	9.73	10
2	9.8	10
3	8.8	9
4	8.87	9
5	7.87	8
6	7.94	8
7	6.94	7
8	7.01	7
9	6.01	6
10	6.08	6
11	5.08	5
12	5.15	5
13	4.15	4
14	4.22	4
15	3.22	3
16	3.29	3
17	2.29	2
18	2.36	2
19	1.36	1
20	1.43	1
21	0.43	0
22	0	1
23	0.43	0
⋮	⋮	⋮
∞	0	0

$$y(n)T = 0.93\, y(n-1)T + 0.5(-1)^n \tag{4.59}$$

The filter response calculated using equation 4.59, with $y(-1)T = 11$, is summarised in table 4.4, and compared with table 4.1 it is seen that the deadband effect has been significantly reduced by the addition of the dither signal.

4.6 OVERFLOW OSCILLATIONS

In section 4.2.3 it was briefly mentioned that in using a fixed-point format for binary numbers, the arithmetic addition of two numbers may produce an overflow in the result. If overflow does occur the digital filter behaves in a non-linear manner, and it has been shown by Ebert *et al.*[19] that under this condition undesirable self-sustained oscillations can exist within the filter.

The input–output characteristic of a two's complement arithmetic adder is shown in figure 4.16a, and it is seen that overflow will occur if the total input to the adder exceeds ± 1. To avoid these undesirable overflow oscillations the adder may be modified to *saturate*, thereby having an input–output characteristic of the form shown in figure 4.16b. Indeed, many digital filters are implemented using *saturating adders*[19] to eliminate overflow oscillations.

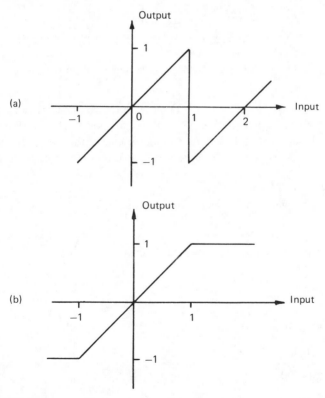

Figure 4.16   (a) Input–output characteristic of a two's complement adder; (b) input–output characteristic of a saturating adder

It is worth noting that in using the normal type of *non-saturating adder* (figure 4.16a) partial sums may be permitted to overflow, as long as their final sum is < 1. By referring to figure 4.17 it is clearly seen why overflows can be temporarily

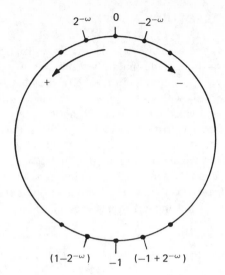

Figure 4.17 Two's complement, fixed-point binary number representation–defined by equation 4.2

tolerated. We see that in using a $(w + 1)$-bit word length (includes sign bit) and the two's complement representation defined by equation 4.2, an overflow will produce a number within the dynamic number range of the $(w + 1)$-bit word, albeit a temporary erroneous result. Example 4.10 illustrates this point.

Example 4.10
Using two's complement arithmetic form the sum of

$$\tfrac{3}{8} + \tfrac{3}{4} + \tfrac{1}{8} - \tfrac{1}{2}$$

and show that temporary overflows are tolerable.

SOLUTION

$$\tfrac{3}{8} = 0*011$$

$$\tfrac{3}{4} = 0*110$$

$$\tfrac{1}{8} = 0*001$$

$$-\tfrac{1}{2} = 1*100$$

First partial sum is

$$+ \begin{array}{l} 0*011 = \tfrac{3}{8} \\ 0*110 = \tfrac{3}{4} \\ \hline 1*001 = -\tfrac{7}{8} \end{array} \quad \text{(overflow has produced erroneous result)}$$

Second partial sum is

$$+ \begin{array}{l} 1*001 = -\frac{7}{8} \\ 0*001 = \frac{1}{8} \\ \hline 1*010 = -\frac{3}{4} \end{array}$$

Third partial sum is

$$+ \begin{array}{l} 1*010 = -\frac{3}{4} \\ 1*100 = -\frac{1}{2} \\ \hline 10*110 = \text{final sum} \end{array}$$

The final sum is a 4-bit word $= +\frac{3}{4}$, which is the correct answer, hence we see that temporary overflows can be tolerated.

4.7 CONCLUDING REMARKS

We have seen in this chapter that the accuracy of a digital filter is governed by the finite word length used in its implementation. Using the analytical techniques presented in this chapter a reasonable estimate of the necessary word length can be made, and if possible a *safety factor* should be used such that three or four bits are added to the estimated word length.

Because of its simplicity fixed-point arithmetic is preferred over floating-point arithmetic in the filter implementation. Furthermore, the analysis of floating-point errors is generally more complicated than fixed-point errors. The readers interested in learning about floating-point arithmetic errors may start their studies by consulting references 20, 21 and 22.

REFERENCES

1. T. J. Terrell, 'A Demonstration of Digital Filter Design, Implementation and Testing', *Int. J. Elect. Engng Educ.*, 14 (1977) 221–36.
2. R. De Mori, S. Rivoira and A. Seria, 'A Special Purpose Computer for Digital Signal Processing', *Trans. on Computing, IEEE*, 24 (1975) 1202–11.
3. M. Yaacob, 'Digital Filtering Using a Microprocessor', M.Sc. Dissertation, University of Manchester, 1977.
4. R. A. Comley and J. E. Brignell, 'Digital Filter Implementation by means of Slave Processors', *Colloquium Digest No. 1977/50, IEE*, (1977) 51–6.
5. A. Peled and B. Liu, 'A New Hardware Realization of Digital Filters', *Trans. Acoustics, Speech and Signal Processing, IEEE*, 22 (1974) 456–62.
6. G. Forsythe and C. B. Moler, *Computer Solution of Linear Algebraic Systems* (Prentice-Hall, Englewood Cliffs, N.J., 1967).
7. R. E. Bogner and A. G. Constantinides, *Introduction to Digital Filtering* (Wiley, New York and London, 1975) Appendix 10c, 179–81.
8. B. Liu, 'Effect of Finite Word Length on the Accuracy of Digital Filters—A Review', *Trans. Circuit Theory, IEEE*, 18 (1971) 670–7.

9. L. R. Rabiner and B. Gold, *Theory and Application of Digital Signal Processing* (Prentice-Hall, Englewood Cliffs, N.J., 1975) chapter 5.
10. A. V. Oppenheim and R. W. Schafer, *Digital Signal Processing* (Prentice-Hall, Englewood Cliffs, N.J., 1975) chapter 9.
11. F. F. Kuo and J. F. Kaiser, *Systems Analysis by Digital Computers* (Wiley, New York, 1966) chapter 7.
12. J. F. Kaiser, 'Some Practical Considerations in the Realization of Linear Digital Filters', *Proceedings of the 3rd Annual Alterton Conference on Circuit and System Theory*, 1965.
13. L. B. Jackson, 'An Analysis of Limit Cycles Due to Multiplication Rounding in Recursive Digital (Sub) Filters', *Proceedings of the 7th Annual Alterton Conference on Circuit and System Theory*, 1969.
14. L. J. Long and T. N. Trick, 'An Absolute Bound on Limit Cycles Due to Roundoff Errors in Digital Filters', *Trans. Audio and Electroacoustics, IEEE*, 21 (1973) 27–30.
15. S. R. Parker and S. F. Hess, 'Limit-cycle Oscillations in Digital Filters', *Trans. Circuit Theory, IEEE*, 18 (1971) 687–97.
16. I. W. Sandberg and J. F. Kaiser, 'A Bound on Limit Cycles in Fixed-Point Implementations of Digital Filters', *Trans. Audio and Electroacoustics, IEEE*, 20 (1972) 110–12.
17. R. B. Blackman, *'Linear Data-Smoothing and Prediction in Theory and Practice'*, (Addison-Wesley, Reading, Mass., 1965).
18. N. Jayant and L. Rabiner, 'The Application of Dither to the Quantization of Speech Signals', *Bell Syst. tech. J.*, 51 (1972) 1293–1304.
19. P. M. Ebert, J. E. Mazo and M. G. Taylor, 'Overflow Oscillations in Digital Filters', *Bell Syst. tech. J.*, 48 (1969) 2999–3020.
20. B. Liu and T. Kaneko, 'Error Analysis of Digital Filters Realized with Floating-Point Arithmetic', *Proc. IEEE*, 57 (1969) 1735–47.
21. I. W. Sandberg, 'Floating-Point Round-Off Accumulation in Digital Filter Realization, *Bell Syst. tech. J.*, 46 (1967) 1775–91.
22. T. Kaneko and B. Liu, 'Accumulation of Round-Off Error in Fast Fourier Transforms', *J. Ass. comput. Mach.*, 17 (1970) 637–54.
23. L. B. Jackson, *Digital Filters and Signal Processing* (Kluwer Academic Publishers, Boston, 1986) chapter 11.

PROBLEMS

4.1 A digital filter has a pulse transfer function $G(Z) = 0.5/(Z^2 - 0.6Z + 0.08)$; determine the average output noise power in

(a) a *direct form* implementation;
(b) *cascade form* of implementation; and
(c) *parallel form* of implementation.

4.2 A digital filter has a pair of complex conjugate poles at $Z = (0.9 \pm j0.3)$, determine

(a) the minimum word length for stability to be maintained, assuming that filter coefficients are truncated; and
(b) the minimum word length required to hold the nominal pole positions to within 0.1 per cent of their desired locations; use the sensitivity to changes in b_2 in evaluating the answer.

4.3 A recursive first-order digital filter is represented by the following linear difference equation

$$y(n)T = x(n)T - 0.95\, y(n-1)T$$

Determine the filter's impulse response under the conditions listed below

(a) $y(-1)T = 0$
(b) $y(-1)T = 10$
(c) $y(-1)T = 11$ $x(n)T = \{10, 0, 0, 0, \ldots 0\}$
(d) $y(-1)T = 12$

Assume that the filter output is rounded to the nearest integer. For this filter operating under condition (a) above, a dither signal is to be added to the right-hand side of the linear difference equation to cause a substantial reduction in the deadband effect. What dither signal will achieve this objective?

5 Practical Implementation of Digital Filters

5.1 INTRODUCTION

Having selected or derived the desired digital filter pulse transfer function, $G(Z)$, (see chapter 2 and chapter 3) and, furthermore, having determined the appropriate processor word length (see chapter 4), then the next step is to undertake the implementation of the filter's linear difference equation corresponding to $G(Z)$. This linear difference equation is used to compute the filter's output $y(n)T$ values, which will be a filtered version of the filter's input $x(n)T$ values (see figure 1.1).

The implementation of the filter's linear difference equation is clearly an essential undertaking, and it is generally achieved in one of two ways. One way to implement the filter is by using *dedicated hardware* to perform the $y(n)T$ computations referred to above. An alternative way to implement the filter is by programming a microcomputer (microprocessor) or a digital signal processor chip to perform the necessary computations. Both methods of implementation are described in the following sections of this chapter.

In an age of rapidly changing digital technology it is prudent, and indeed probably more beneficial, to study and understand the techniques involved in the filter implementation, rather than try to swallow and digest complex detail concerning the processor hardware and software. With this point in mind it was decided to confine the majority of this chapter's hardware/software detail to the appendixes at the end of the chapter.

5.2 IMPLEMENTATION USING DEDICATED HARDWARE

A digital filter which is suitable for implementation using dedicated hardware will generally have certain features which recommend it for this method of realisation.

The main features are

(1) It is required to operate in *real time*.

(2) It is required to perform a well defined filtering operation, that is, the a_i and b_i coefficients in the linear difference equation are held constant for a particular filter configuration.

(3) It will normally use fixed-point binary arithmetic.

(4) In replacing an existing analogue filter it is required that the digital filter be superior as far as cost and/or performance is concerned.

(5) It is required that the digital filter be a relatively simple self-contained autonomous digital system.

In addition to the above listed attributes, the digital filter will have a feature which is common to many types of digital signal processor, namely, it will have to perform *array multiplications*. Such array multiplications can be expressed in the form

$$A = \sum_{k=1}^{I} d_k B_k \tag{5.1}$$

where d_k is the set of constant filter coefficients, and B_k are input or output sampled-data values $[x(n)T; y(n)T]$, or they are intermediate values $[x(n-1)T; y(n-1)T;$ etc.] . For example, the linear difference equation corresponding to the digital filter pulse transfer function given in equation 2.67 is

$$y(n)T = 0.3333x(n)T + 0.0330x(n-1)T - 0.1824y(n-1)T$$
$$+ 0.1126y(n-2)T$$

which serves to illustrate a typical practical application of equation 5.1.

We saw in section 4.2.1 that a signed decimal fraction may be represented by a $(w+1)$-bit binary number in two's complement form, as given by equation 4.2. If we now assume that the filter's sampled-data values, $x(n)T$, $x(n-1)T$, . . ., $y(n)T$, $y(n-1)$, . . ., are restricted in amplitude, such that $|B_k| < 1$, then the B_k values may be represented as a $(w+1)$-bit binary number in two's complement form as defined by equation 4.2. Using the notation of equation 4.2, the array multiplications given by equation 5.1 may be rewritten as

$$A = \sum_{k=1}^{I} \left[d_k \left(-C_0 + \sum_{j=1}^{w} C_j 2^{-j} \right)_k \right]$$

$$= \sum_{k=1}^{I} \left[-d_k \left(C_0 \right)_k + d_k \sum_{j=1}^{w} \left(C_j 2^{-j} \right)_k \right]$$

$$= - \sum_{k=1}^{I} d_k \left(C_0 \right)_k + \left[\sum_{k=1}^{I} d_k \sum_{j=1}^{w} \left(C_j 2^{-j} \right)_k \right] \tag{5.2}$$

Now interchanging the order of summation we obtain

$$A = - \sum_{k=1}^{I} d_k \left(C_0\right)_k + \left[\sum_{j=1}^{w} 2^{-j} \sum_{k=1}^{I} d_k \left(C_j\right)_k\right] \qquad (5.3)$$

Since both $(C_0)_k$ and $(C_j)_k$ are either 0 or 1, then correspondingly $\sum_{k=1}^{I} d_k (C_0)_k$ and $\sum_{k=1}^{I} d_k (C_j)_k$ will each have 2^I possible values. For example, table 5.1 lists the four possible values occurring when $I = 2$. In practice, in the implementation of the filter the 2^I values are normally stored in a read only memory (ROM) (see

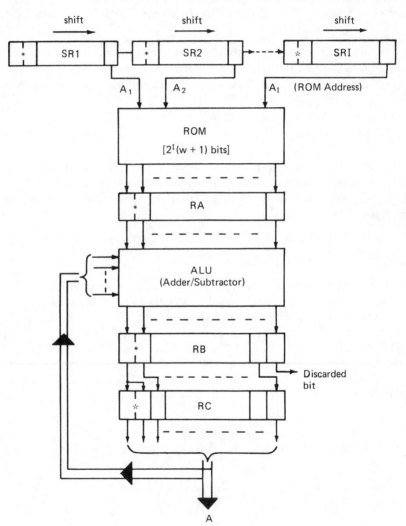

Figure 5.1 Block diagram of hardware for computing the array multiplication

appendix 5.1 for a description of the ROM). Thus using the *addressed stored values* held in ROM the computation of A (equation 5.3) may be simply stated to be

$$A = - \begin{bmatrix} \text{addressed} \\ \text{stored value} \\ (j = 0) \end{bmatrix} + \begin{bmatrix} \sum_{j=1}^{w} 2^{-j} \begin{pmatrix} \text{addressed} \\ \text{stored value} \end{pmatrix} \end{bmatrix} \qquad (5.4)$$

Table 5.1 Addressed stored values for $I = 2$

C_{01}/C_{j1}	C_{02}/C_{j2}	$\left[\sum_{k=1}^{2} d_k (C_0)_k\right] \Big/ \left[\sum_{k=1}^{2} d_k (C_j)_k\right]$
0	0	0
1	0	d_1
0	1	d_2
1	1	$d_1 + d_2$

The block schematic diagram shown in figure 5.1 represents the hardware of a non-recursive digital system which may be used to compute the value of the array multiplication, A. Referring to figure 5.1 the computation of A (equation 5.4) is achieved as outlined below.

(1) Initially the $(w + 1)$-bit registers, RA, RB, RC, SR1, SR2,...., SRI are cleared.
(2) The $(w + 1)$-bit binary representation of $x(n)T$ is of the form

$$\left(-C_0 + \sum_{j=1}^{w} C_j 2^{-j} \right)_k$$

where $k = 1, 2, \ldots, I$, which is the filter's sampled-data input value, and it is loaded (least significant bit leading) into the $(w + 1)$-bit shift register, SR1 (shift registers are described in appendix 5.1).
(3) The least significant bit of each shift register $(A_1, A_2, \ldots, A_I)$ is used to address the ROM, thereby accessing the *addressed stored value* referred to in equation 5.4. This *addressed stored value* is then loaded into the $(w + 1)$-bit register RA.
(4) The sum RC + RA is formed using an arithmetic logic unit (ALU) and the result is held in the $(w + 1)$-bit register RB. That is, RB = RC + RA, thereby producing the first partial result (see appendix 5.1 for a description of the ALU).
(5) The contents of register RB are copied into register RC, as shown in figure 5.1. Thus the contents of register RB are effectively shifted one place to the right and held in register RC to produce a multiplication by 0.5 (the 2^{-j} term in equation 5.4).

(6) The contents of shift registers SR1, SR2, ..., SRI are shifted one place right.

Steps (3) to (6) inclusive are repeated again w times, and the last time through step (4) the difference RC − RA is formed (corresponds to $j = 0$: the sign bit). Thus at the end of $(w + 1)$ iterations of steps (3) to (6) inclusive, the value of A is held in register RB. At this point in the system operation register RC is cleared ready for the next computation. The binary representation of A may then be transferred in parallel to a *latch register* (see appendix 5.1) for subsequent D/A conversion, thereby forming the filter's sampled-data value, $y^*(t)$. The next $x(n)T$ filter input is obtained via the A/D converter, which brings us back to executing step (2) in the sequence. The new value of A is then obtained by repeating the sequence of steps (3) to (6) through $(w + 1)$ iterations as outlined above. The following worked example demonstrates how the system operates.

Example 5.1
Using the system shown in figure 5.1 (described above) show how the hardware is used to compute the $y(3)T$ filter output value; the linear difference equation of the filter is $y(n)T = 0.75x(n)T - 0.5x(n - 1)T$. Assume that $x(n)T = 0$ for $n < 0$, and $x(n)T = \{0.141, 0.181, 0.28125, 0.5, 0.3, ...\}$ for $n = 0, 1, 2, 3, ...$ The processor has a word length of 6 bits. Comment on the result.

SOLUTION
$$y(3)T = 0.75x(3)T - 0.5x(2)T$$

Therefore

$$y(3)T = (0.75 \times 0.5) - (0.5 \times 0.28125)$$

that is

$$y(3)T = 0.234375 = 0*001111$$

For the hardware $w + 1 = 6$ (6 bit word length), therefore the sampled-data values and filter coefficients are

$$0.5 = 0*10000 = x(3)T$$

$$0.28125 = 0*01001 = x(2)T$$

$$0.75 = 0*11000 = \text{coefficient } d_1$$

$$-0.5 = 1*10000 = \text{coefficient } d_2$$

$$0.75 - 0.5 = 0*01000 = d_1 + d_2$$

According to table 5.1 and figure 5.1 the *ROM map* in this example is

ROM Contents	*ROM address:* A_1 A_2
0*00000	0 0
0*11000	1 0
1*10000	0 1
0*01000	1 1

The first iteration progresses as follows.

Step (1). RA = RB = RC = SR1 = SR2 = 0*00000, that is, these registers are initially cleared.

Step (2). SR1 is loaded with the $x(3)T$ value, and SR2 contains the $x(2)T$ value from the previous iteration, that is, SR1 = 0*10000 and SR2 = 0*01001.

Step (3). ROM address bit, A_1, is equal to the least significant bit of SR1, and ROM address bit, A_2, is equal to the least significant bit of SR2. That is, $A_1 = 0$ and $A_2 = 1$. The corresponding addressed stored value in the ROM is loaded into RA. That is, RA = 1*10000. At this step RA is the only register that changes its contents.

Step (4). The ALU adds RA to RC, and the result is loaded in RB. That is

$$RA = 1*10000 \quad \text{(see step (3))}$$
$$RC = 0*00000 \quad \text{(see step (1))}$$
$$\overline{}$$
$$RA + RC = 1*10000 = RB$$

At this step RB is the only register that changes its contents.

Step (5). The contents of RB are copied into RC using a hardwired *skew parallel connection* (right shift by 1 bit). However, it is important to realise that in this step the sign bit of RB is copied to yield the sign bit and most significant bit of RC, as shown in figure 5.1.

Step (6). Shift registers SR1 and SR2 are shifted one place right. That is

$$SR1 = 0*01000$$
$$SR2 = 0*00100$$

Note that the least significant bit of SR1 is shifted one place right to yield the sign bit of SR2.

At this point the first iteration has been completed. Similarly the remaining iterations are executed by cycling through steps (3) to (6), inclusive, another five times. Table 5.2 summarises the register contents at the start [step (3)] and finish [step (6)] of each iteration.

At the end of six $(w + 1)$ iterations the value of $y(3)T$ is in register RB, that is $y(3)T$ = RB = 0*00111 = 0.21875. Compared with the precise value of $y(3)T$ (0.234375) we see that an error in the calculated value of $y(3)T$ exists due to the limited word length used in the filter implementation (see chapter 4).

In addition to the form of error observed in the above example it is possible that arithmetic overflow will also occur, thereby producing another source of error. For example, if the digital filter in example 5.1 is used with the sampled-data input signal: $x(n)T = 0$ for $n < 0$ and $x(n)T = \{0.75, 0.5, 0.25, 0, \ldots, 0\}$ for $n = 0, 1, 2, 3, \ldots$, then the calculation of $y(0)T = 0.75(0.75) = 0.5625$, would progress as summarised in table 5.3, and we see from the table that arithmetic overflow is present at step (6) in the fifth iteration, having occurred as the result of forming the sum RB = RA + RC at step (4) in this iteration. Obviously the final

Table 5.2 Summary of register contents for example 5.1

Iteration No.	Step Executed	SR1 Content	SR2 Content	ROM Address A_1	ROM Address A_2	RA Content	RB Content	RC Content
Initialisation	(1)	0*00000 x(3)T	0*00000	0	0	0*00000	0*00000	0*00000
	(2)	0*10000 x(3)T	0*01001 x(2)T	0	1	0*00000	0*00000	0*00000
1	(3)	0*10000	0*01001	0	1	1*10000	0*00000	0*00000
	(6)	0*01000	0*00100	0	0	1*10000	1*10000	1*11000
2	(3)	0*01000	0*00100	0	0	0*00000	1*10000	1*11000
	(6)	0*00100	0*00010	0	0	0*00000	1*11000	1*11100
3	(3)	0*00100	0*00010	0	0	0*00000	1*11000	1*11100
	(6)	0*00010	0*00001	0	1	0*00000	1*11100	1*11110
4	(3)	0*00010	0*00001	0	1	1*10000	1*11100	1*11110
	(6)	0*00001	0*00000	1	0	1*10000	1*01110	1*10111
5	(3)	0*00001	0*00000	1	0	0*11000	1*01110	1*10111
	(6)	0*00000	1*00000	0	0	0*11000	0*01111	0*00111
6	(3)	0*00000	1*00000	0	0	0*00000	0*01111	0*00111
	(6)	0*00000	0*10000	0	0	0*00000	0*00111	0*00011

$y(3)T = 0.21875$

Table 5.3 Summary of register contents to demonstrate occurrence of an overflow

Iteration No.	Step Executed	SR1 Content	SR2 Content	ROM Address A_1	A_2	RA Content	RB Content	RC Content
Initialisation	(1)	0*00000	0*00000	0	0	0*00000	0*00000	0*00000
	(2)	0*11000 $x(0)T$	0*00000 $x(-1)T$	0	0	0*00000	0*00000	0*00000
1	(3)	0*11000	0*00000	0	0	0*00000	0*00000	0*00000
	(6)	0*01100	0*00000	0	0	0*00000	0*00000	0*00000
2	(3)	0*01100	0*00000	0	0	0*00000	0*00000	0*00000
	(6)	0*00110	0*00000	0	0	0*00000	0*00000	0*00000
3	(3)	0*00110	0*00000	0	0	0*00000	0*00000	0*00000
	(6)	0*00011	0*00000	1	0	0*00000	0*00000	0*00000
4	(3)	0*00011	0*00000	1	0	0*11000	0*00000	0*00000
	(6)	0*00001	1*00000	1	0	0*11000	0*11000	0*01100
5	(3)	0*00001	1*00000	1	0	0*11000	0*11000	0*01100
	(6)	0*00000	1*10000	0	0	0*11000	1*00100†	1*10010
6	(3)	0*00000	1*10000	3	0	0*00000	1*00100	1*10010
	(6)	0*00000	1*10000	0	0	0*00000	1*10010	0*00000

$$y(0)T = -0.875$$

†Overflow

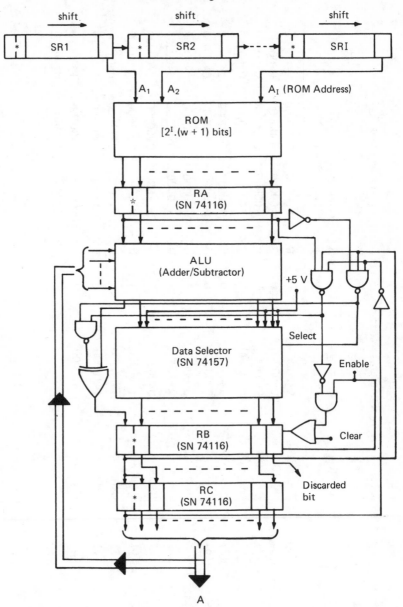

Figure 5.2 Block diagram of modified hardware incorporating a saturating output on the ALU

answer $y(0)T = -0.875$ is an erroneous result due to the arithmetic overflow.

To reduce the error produced by arithmetic overflow the system shown in figure 5.1 may be modified to the form shown in figure 5.2, thereby extending the ALU to effectively incorporate a saturating output when overflow occurs. The basis of the saturating process is best illustrated by two simple examples, as follows. Let

$$RC = -0.75 = 1*010$$

and let

$$RA = -0.5 \ = 1*100$$

Summing we obtain

$$1 \ 0*110 = RB = RC + RA$$

We see that overflow has occurred, and consequently to reduce the computation error register RB must be altered (made to saturate) to equal -1 (1*000). In this case this is achieved by inverting the sign-bit, C_0 (RB), and clearing all other bits in register RB to logical 0. Similarly, we could have had the following sum

$$RC = 0.75 = 0*110$$
$$RA = 0.5 \ = 0*100$$
$$1*010 = RB = RC + RA$$

Again we see that an error has occurred, and in this case to reduce the computation error, register RB must be altered to equal $(1 - 2^{-w})$, that is, in this example RB will be set to 0*111. This is achieved by inverting the sign-bit, C_0 (RB), and setting all the other bits in register RB to logical 1.

By inspecting the two examples above we may deduce a logical expression which describes the event of an overflow or error occurrence. That is, if we signify the occurrence of an overflow or error (O/E) by logical 1, then we may write

$$O/E = [\overline{C_0 \ (RA)} \cdot \overline{C_0 \ (RC)} \cdot C_0 \ (RB)] + [C_0 \ (RA) \cdot \overline{C_0 \ (RC)} \cdot C_0 \ (RB)] \quad (5.5)$$

Table 5.4 gives the truth table corresponding to equation 5.5. Figure 5.2 shows the logic used to achieve overflow or error detection and modification of the contents of register RB. That is, when an overflow or error is detected the logic saturates register RB to the appropriate value shown in table 5.4.

Table 5.4 Truth-table corresponding to equation 5.5

C_0 (RA)	C_0 (RC)	C_0 (RB)	O/E
0	0	0	0
0	0	1	1 (saturate to (0*111. . .1)
0	1	0	0
0	1	1	0
1	0	0	0
1	0	1	0
1	1	0	1 (saturate to (1*000. . .0)
1	1	1	0

Table 5.5 Summary of register contents to demonstrate the effect of incorporating a saturating output

Iteration No.	Step Executed	SR1 Content	SR2 Content	ROM Address	RA Content	RB Content	RC Content
5	(3) see table 5.3	0*00001	1*00000	1	0*11000	0*11000	0*01100
	(6)	0*00000	1*10000	0	0*11000	1*00100† 0*11111	0*01111
6	(3)	0*00000	1*10000	0	0*00000	0*11111	0*01111
	(6)	0*00000	0*11000	0	0*00000	0*01111	0*00000

$$A = y(0)T = 0.46875$$

†Saturate

To see the value of using a saturating adder in the filter implementation we can take the results which are summarised in table 5.3, and at the point at which overflow occurs [step (4), fifth iteration] we can modify the content of register RB so that it saturates to the appropriate value. This action is summarised in table 5.5, and we see that the final result is equal to 0.46875, which is much closer to the ideal (0.5626) when compared with the result obtained previously (−0.875) in table 5.3.

We have seen that the method of implementation described above involves I multiplications and $(I - 1)$ additions in the implementation of the *array multiplication* defined by equation 5.1. Furthermore, we have seen that the ROM must have at least 2^I addressable memory locations, each location having a word length of $(w + 1)$-bits. It is possible to reduce the ROM storage requirement in the filter implementation by allowing a corresponding increase in the number of additions. For example, if we let $I = UV$, then equation 5.1 may be rewritten thus

$$A = \sum_{j=1}^{V} W_j \qquad (5.6)$$

where

$$W_j = \sum_{k=[(j-1)U]+1}^{Uj} d_k B_k \qquad (5.7)$$

Equation 5.7 is the same form as equation 5.1, and therefore the partial sum W_j may be implemented using a hardware configuration similar to that shown schematically in figure 5.1. The final sum, A, is obtained by adding together the V partial sums defined by equation 5.6.

Example 5.2 demonstrates how the ROM storage requirements may be considerably reduced, with a corresponding increase in the number of additions in the filter implementation.

Example 5.2
If I is 14, calculate: (a) the ROM storage requirements for the implementation method shown schematically in figure 5.1; and (b) the ROM storage requirements and number of extra additions required when the value of A is calculated using the method defined by equations 5.6 and 5.7. Comment on the results.

SOLUTION
(a) The ROM storage requirements are $2^I = 2^{14} = 16384$ locations [words of length $(w + 1)$-bits]. Number of additions = $(I - 1) = 13$.
(b) $I = 14$, let $V = 2$, therefore $U = 7$, $(UV = I)$. In this case the ROM storage requirements are

$$V2^U = 2.2^7 = 256 \text{ locations}$$

Number of additions = $(I - 1) + (V - 1) = 14$
We see that in (b) only 1/64 of the ROM in (a) is required for the filter implementation. In implementing (b) instead of (a) we would achieve a substantial

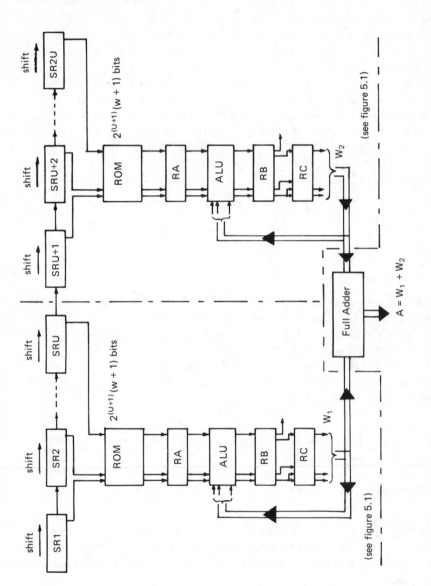

Figure 5.3 Block diagram of hardware used to reduce ROM storage requirements

saving in ROM storage requirements, and the resulting penalty would be: (i) a slight loss in processor speed due to having to perform one more addition; and (ii) some additional hardware must be provided to handle the formation of the partial sums W_1 and W_2; one form of implementation is shown schematically in figure 5.3.

The basic form of implementation considered thus far is seen to be suitable for the implementation of non-recursive digital filters, because in this case partial results need not be stored for subsequent transfer into the appropriate shift registers. Indeed, for the non-recursive case partial results exist in the appropriate shift register due to the serial shifting of the data through $(w + 1)$ iterations. We therefore see that a *bit serial operation* offers the advantage of having a hardware implementation which is relatively simple, albeit mainly restricted to non-recursive filters. However, the *bit serial* form of implementation does mean that the speed of system operation will be relatively slow. In contrast a bit parallel operating system offers the fastest speed of computation, but the price paid is a corresponding increase in hardware complexity. Furthermore, a *bit parallel operating system* is suitable for the implementation of recursive digital filters. The following worked example demonstrates how a simple first-order recursive digital filter may be implemented as a *bit parallel operating system*.

Example 5.3
The linear difference equation of a recursive digital filter is $y(n)T = 0.75x(n)T - 0.5y(n-1)T$. Assume $y(-1)T = 0$ and that $x(n)T = 0$ for $n < 0$ and $x(n)T = 0.5$ for all $n \geqslant 0$, where n is an integer. The processor has a word length of 6 bits. Using a *bit parallel* form of implementation show how the system calculates the $y(2)T$ filter output value.

SOLUTION
Ideally, using the given linear difference equation we obtain

$$y(0)T = 0.75(0.5) - 0.5(0) = 0.375$$

$$y(1)T = 0.75(0.5) - 0.5(0.375) = 0.1875$$

$$y(2)T = 0.75(0.5) - 0.5(0.1875) = 0.28125$$

The *bit parallel computation* of A may be achieved by adding together simultaneously, through an *adder tree*, addressed stored values held in $(w + 1)$ ROMs (see figure 5.4). The filter coefficients in this example are identical to the filter coefficients used in example 5.1, therefore the *basic ROM map* (BRM) in this example will be the same as that used in example 5.1. That is, the basic ROM map for this example is

ROM contents	ROM address: A_1 A_2
0*00000	0 0
0*11000	1 0
1*10000	0 1
0*01000	1 1

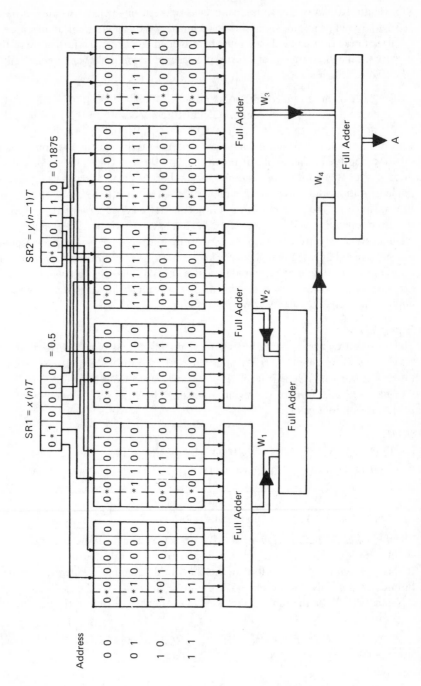

Figure 5.4 Block diagram of hardware for a bit-parallel operating system

Referring to equation 5.4 we see that the ROM addressed by the sign-bits of $x(n)T$ and $y(n - 1)T$ must contain *negative addressed stored values*, that is the two's complement of the basic ROM map. Also we see that the ROM addressed by the two most significant bits of $x(n)T$ and $y(n -- 1)T$ contains *addressed stored values* of the basic ROM map multiplied by 2^{-1}. Therefore we see that as we move right, bit by bit, along shift registers SR1 and SR2, the ROM addressed by the 2^{-v} pair of bits, where $1 \leqslant v \leqslant w$, will contain the addressed stored values of the basic ROM map multiplied by 2^{-v}, as shown in figure 5.4.

Now referring to figure 5.4 the computation of $y(2)T$ proceeds as follows

      ROM 1 output = 0*00000
      ROM 2 output = 0*01100
      ROM 3 output = 0*00000
      ROM 4 output = 1*11110
      ROM 5 output = 1*11111
      ROM 6 output = 0*00000
      Full adder output, W_1 = ROM 1 output + ROM 2 output = 0*01100
      Full adder output, W_2 = ROM 3 output + ROM 4 output = 1*11110
      Full adder output, W_3 = ROM 5 output + ROM 6 output = 1*11111
      Full adder output, W_4 = $W_1 + W_2$ = 1 ⫶ 0*01010
      Full adder output, A = $W_3 + W_4$ = 1 ⫶ 0*01001 = 0.28125

The techniques so far presented in this chapter have been concerned with two useful methods of implementing the *array multiplication* which is inherent in digital signal processors (digital filters). Obviously the *bit parallel* form of implementation has the fastest possible processing speed, and in contrast the *bit serial* form of implementation has the slowest possible processing speed. Generally the price to be paid for achieving an increase in processing speed is a corresponding increase in hardware. Therefore, in using dedicated hardware for the implementation of digital filters, the filter-designer will have to estimate the 'breakover point' between processor speed and cost of hardware. In some practical cases it may be necessary to trade off a loss in processor speed for a corresponding reduction in cost of hardware. Thus in using dedicated hardware the filter-designer will have to be satisfied that firstly the processor speed (sampling frequency) is high enough to achieve the desired filtering operation, and secondly that the method of implementation is cost effective.

A further consideration is the possibility of using a microcomputer for the filter implementation. This form of implementation is suitable because in this case the cost of hardware is now comparable with filters implemented using dedicated hardware. Clearly there are some digital filters which readily lend themselves to implementation on a general purpose digital microcomputer—for example, in the case where the filter has to be adaptive, thereby necessitating a change in the filter coefficients and a corresponding change in the computational process. Therefore it is not surprising that considerable effort has been directed towards the design and implementation of general purpose digital signal processors.

The next section provides an introduction to the implementation of digital filters using general purpose microprocessor-based systems.

5.3 IMPLEMENTATION USING MICROPROCESSORS

We have already seen that second-order sections play an important role in digital filter implementations, and we know that this type of filter may be represented by a linear difference equation of the form

$$y(n)T = [a_0 x(n)T + a_1 x(n-1)T + a_2 x(n-2)T + b_1 y(n-1)T$$

$$+ b_2 y(n-2)T] \qquad (5.8)$$

which may be implemented using a suitable microprocessor.

Once the algorithm (linear difference equation) describing the filter has been derived or defined, the first step in the programming process is to decide how the data and computer instructions are to be organised. In order to achieve this objective the programmer may use a diagram, commonly referred to as a *flowchart*, to formulate the logical sequence of steps involved in implementing the filter.

5.3.1 Flowcharts

A flowchart is made up of several standard symbols, which are connected together by straight lines. Arrow-heads are drawn on the connecting lines to indicate the *direction of flow* in the program. A selection of commonly used flowchart symbols is shown in figure 5.5.

The start and end of the flowchart is indicated by the *terminal symbol* shown in figure 5.5a, and obviously it is connected to the flowchart by only one line. The *rhomboid symbol* shown in figure 5.5b is used to show a specific operation by an input or output device, for example, data input from a Teletype console. The *rectangle symbol* shown in figure 5.5c is used to indicate that a specific action is to be taken. A statement inside the rectangle specifies the action, and this may be stated in plain English, or alternatively it could be any logical or algebraic expression. For example, we could write the following statement inside the rectangle

> Add one to store location 5 and then copy the contents of store location 5 into store location 23.

The above statement can be conveniently abbreviated using a suitable notation, as follows

$$S(23) = S(5) + 1$$

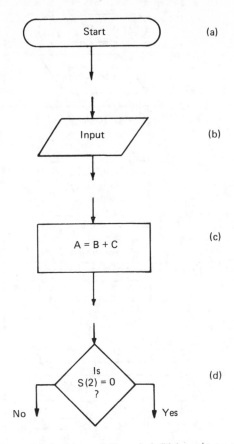

Figure 5.5 Flowchart symbols: (a) terminal symbol; (b) input/output symbol; (c) action symbol; (d) decision symbol

where the notation $S(N)$ is interpreted to mean the content of store location N.

The *diamond symbol* shown in figure 5.5d is used to indicate the point in a program at which a decision has to be made. For example, it may be necessary at some point in the program to know whether or not the contents of store location 2 are equal to zero (see figure 5.5d); obviously the answer to this question is either no or yes. Clearly the program will *branch* along one of two possible paths, depending on the decision made. A typical flowchart is given in figure 5.6, and it outlines the basic steps involved in implementing the digital filter defined by equation 5.8.

The next step in the programming process is to translate the flowchart into a computer program. The implementation (programming) of equation 5.8 may be achieved using either a high-level language, such as FORTRAN, PASCAL or BASIC, or a machine-code language.

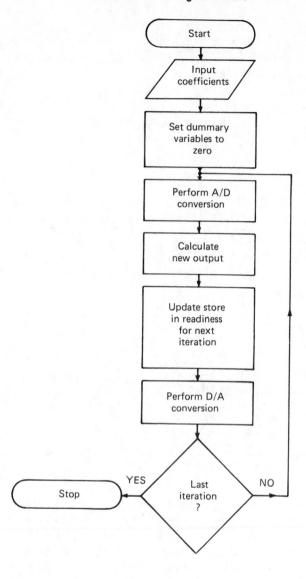

Figure 5.6 Flowchart for the digital filter defined by equation 5.8,

In using a high-level language for digital filtering an inherent disadvantage exists, namely the relatively long computation time involved in dealing with a single iteration of the program.

For practical purposes, since a single iterative loop contains one A/D conversion and one D/A conversion, the time estimate obtained is taken to correspond to the sampling period T. Typically from a test using 1000 iterations, it is estimated that T was approximately 130 ms for a given high-level language program. Therefore, in this case, with due regard being paid to the sampling theorem, the highest frequency signal that can be faithfully represented by a sample-set is 3.85 Hz, which will be too restrictive in many applications. Consequently to considerably reduce the sampling period, T, it will be necessary to use an equivalent machine-code program.

The basic concept of programming minicomputers or microcomputers is generally the same in either case, and for this reason—plus the fact that microprocessors are becoming increasingly important in digital signal processing—a microprocessor (the Motorola M6802) will be used to show how machine-code programming is used for the implementation of a digital filter.

5.3.2 Some Pertinent M6802 Microprocessor System Concepts

Microprocessor system organisation is similar to that used in many digital computers, as indicated in figure 5.7.

The read-only memory (ROM), which is described in appendix 5.1, is a *non-volatile store* which is used to hold the sequence of program instructions. The random access memory (RAM, see appendix 5.1), which is a read–write (R/W) store is *volatile*, and it is provided to hold the results of arithmetic and/or logical operations, or variable data derived via the input/output (I/O) interface. In a

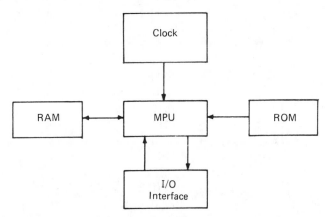

Figure 5.7   Block diagram of a basic microprocessor system

digital filter the I/O interface is used to connect the A/D and D/A converters to the microprocessor unit (MPU).

The word length used in a microprocessor system is an important characteristic because it determines the accuracy of data, processor speed and memory require-ments and cost. None of these factors can be ignored when using a microprocessor for implementing a digital filter.

A word is sometimes conveniently divided into lengths of 8 bits, which is normally referred to as a *byte*. In memory, each store location stores one byte. To enable these bytes to be *accessed* (read or write operation) each store location has an address; a single byte can address any one of 256 memory locations; however, a double byte (16 bits) can address any one of 65 536 memory locations. The byte stored at a particular location is either data, such as a sampled-data input value, or it is an instruction, which will cause the microprocessor to execute a particular operation (see appendix 5.3). The microprocessor will only execute one operation at a time, and consequently the task at hand must be programmed as a series of sequential operations (instructions).

Table 5.6 Code conversion table

Decimal	Binary	Hexadecimal
0	0000	0
1	0001	1
2	0010	2
3	0011	3
4	0100	4
5	0101	5
6	0110	6
7	0111	7
8	1000	8
9	1001	9
10	1010	A
11	1011	B
12	1100	C
13	1101	D
14	1110	E
15	1111	F
16	10000	10

In microprocessor systems the *hexadecimal* (radix 16) code[1] is commonly used, a 4-bit binary number being represented by an hexadecimal character (see the codes listed in table 5.6). A byte, therefore, would be represented by two hexadecimal characters. For example, the binary number 01111011, which we assume is in two's complement form, may be converted to hexadecimal as follows binary   0111 | 1011   (partition into two 4-bit numbers)

hexadecimal   7 | B

that is

$$(01111011)_2 = (7B)_{16} = (123)_{10}$$

Similarly

$$(11111011)_2 = (FB)_{16} = (-5)_{10}$$

It follows that data, instructions and memory addresses may be conveniently represented using hexadecimal notation. Thus we see that all information used by a microprocessor may be represented as hexadecimal codes.

A program instruction may be one, two or three bytes in length. A one-byte instruction would consist of an *op-code* (operation code) only. For example, the Motorola M6802 microprocessor[2] has two accumulator registers, A and B, and op-code $=(16)_{16}$ signifies that the contents of accumulator A will be copied into accumulator B when this instruction is executed. Thus we see that the op-code instructs the microprocessor in what it must do. A two-byte instruction consists of the one byte op-code followed by a one-byte *operand*, the operand being the memory address of the data pertinent to the op-code. The one-byte operand can only have a numerical address in the range $(0$ to $255)_{10}$. A two-byte instruction is referred to as a *direct addressing mode instruction*. For example, op-code = $(DB)_{16}$ signifies that the contents of accumulator B will be added to the contents of the specified direct address, the result being placed in accumulator B. A three-byte instruction consists of the one-byte op-code followed by a two-byte operand. The second byte contains the most significant (highest) 8 bits of the address, and the third byte contains the least significant (lowest) 8 bits of the address. Clearly, in comparison with the direct addressing mode, the use of a 16-bit address will extend the range of accessible memory locations and consequently a three-byte instruction is referred to as an *extended addressing mode instruction*. For example, op-code = $(FB)_{16}$ signifies that the contents of accumulator B will be added to the contents of the specified extended address, the result being placed in accumulator B.

Another form of two-byte instruction is available, known as an *immediate addressing mode instruction*. In this case the op-code is immediately followed by the actual value of the operand. For example, op-code = $(CB)_{16}$ followed by $(2C)_{16}$, signifies that the hexadecimal number 2C will be added to the contents of accumulator B, the result being placed in accumulator B.

It is appropriate at this point to look at the basic programming model of the Motorola M6802 MPU (see figure 5.8). The two 8-bit accumulators, A and B, are generally associated with data movement around the system. They frequently contain one of the operands and the result for a number of instructions. The *program counter* (PC) is 16 bits long, and it holds the address of the next instruction to be executed. The *index register* (IX) is 16 bits long, and it is used to hold

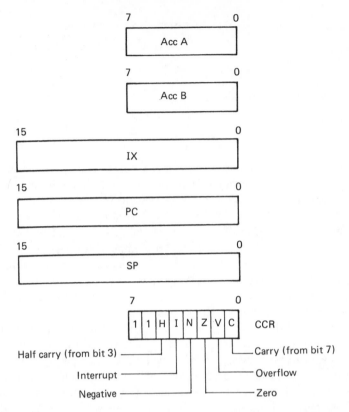

Figure 5.8 Basic programming model of the M6802 (by kind permission of Motorola Semiconductors)

a memory address. When the index register is used a two-byte instruction is employed; the first byte is the op-code, and the second byte is the number which is added to the contents of the index register to form the memory address. This type of instruction is known as an *indexed addressing mode instruction*. For example, op-code = $(EB)_{16}$ followed by $(03)_{16}$, signifies that the contents of accumulator B will be added to the contents of the memory location given by PC = IX + 03, the result being placed in accumulator B.

The *condition code register* (CCR) consists of six flip-flops which 'flag' information about the last operation performed by the MPU. For example, the C bit will 'set' if there was a *carry* from the most significant bit of the result of an addition operation, otherwise it is 'cleared'. The state of a particular *flag* (CCR flip-flop) may be usefully employed to determine the outcome of a *conditional branch* in a program. For example, if a *carry* (C) has been produced we may want the microprocessor to halt further operation otherwise it may continue with the next instruction (see figure 5.9). To achieve this objective we may use the *branch if carry clear instruction*, $(24)_{16}$, which is a *relative addressing mode instruction*.

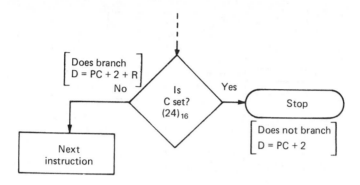

Figure 5.9 Flowchart showing a typical BRANCH operation

A relative addressing mode instruction is a two-byte instruction. The first byte is the op-code (conditional branch), and the second byte is a two's complement, 8-bit, binary number, which acts as an *offset*, R. The relationship between the relative address and the absolute destination address of the branch instruction is given by

$$D = (PC + 2) + R \qquad (5.9)$$

where D is the absolute destination address, PC is the address of the first byte of the branch instruction, and R is the offset. If the program does not branch, then

$$D = (PC + 2) \qquad (5.10)$$

In equation 5.9 the destination address must be within the range specified by

$$[(PC + 2) + 127] \geqslant D \geqslant [(PC + 2) - 128] \qquad (5.11)$$

However, if it is desired to transfer control outside the range specified in equation 5.11, this can be achieved using an *unconditional jump instruction* or *jump to subroutine instruction*.

A section of RAM is used for last-in, first-out (LIFO) operations so that successive bytes may be 'stacked' one after the other. This part of memory is generally referred to as the *stack*. The 16-bit stack pointer (SP) (see figure 5.8) holds the memory address of the next vacant stack location. An accumulator byte may be stored on the stack using a PSH (push) instruction, and the last byte stored on the stack may be retrieved (brought back into an accumulator) using a PUL (pull) instruction. The stack is used in handling *subroutines* (see figure 5.10) and interrupts. For example, when a *jump to subroutine* (JSR) instruction is executed the *return address* is automatically saved on the stack. This return address corresponds to the next instruction after the JSR instruction. A subroutine must be terminated by a *return from subroutine* (RTS) instruction, and after its execution the return address is pulled off the stack and it is placed in the program counter, and thus the next instruction in the main program will be executed.

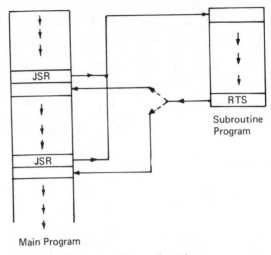

Figure 5.10 Program organisation for handling a subroutine

One of the main features of a microprocessor which make it suitable for digital filtering applications is its ability to handle interrupts. An interrupt signal will be generated by a peripheral device, which in the case of a digital filtering system will be the A/D converter. On receipt of an *interrupt request* (A/D conversion complete) the microprocessor's *status* (contents of the microprocessor's internal registers) are stored on the stack, and they are restored to the microprocessor on completion of the *interrupt service routine* (ISR). Thus basically the interrupt signal will cause the microprocessor to suspend execution of its main program and execute the interrupt service routine, and at the end of this routine the *return*

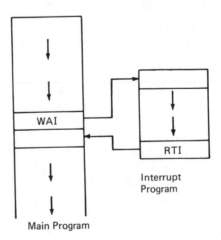

Figure 5.11 Program organisation for handling an interrupt routine

from interrupt (RTI) instruction is encountered and executed, thereby enabling the resumption of the main program (see figure 5.11). Generally, in the case of a digital filter implementation the ISR would be used mainly to evaluate the filter's linear difference equation–this is demonstrated later in this chapter.

Another important and essential consideration in digital filter implementation is the *interface* between the MPU and the A/D and D/A converters. The input/output of data between the MPU and the converters may be achieved under program control, through programmable interface adapters (PIAs) (see figure 5.12). Motorola's family of microprocessor components contains the MC6821 PIA,[2] and this will now be described.

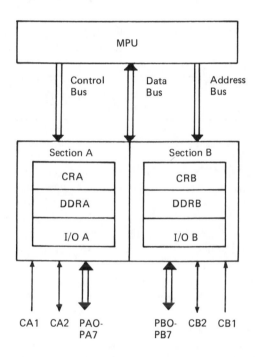

Figure 5.12 Block diagram of the MC6821 PIA

Referring to figure 5.12 we see that the PIA is divided into two sections, A and B. Each section has three 8-bit registers: a control register, a data direction register, and an input/output (I/O) register. The A section has eight data I/O lines, designated as PA0, PA1, PA2, . . ., PA7, and it has two control lines, designated as CA1 and CA2. Similarly, the B section has eight data I/O lines, designated as PB0, PB1, PB2, . . ., PB7, and it has two control lines, designated as CB1 and CB2.

This PIA is a memory-mapped device, that is, its six internal registers are effectively treated as addressable RAM locations. In the remainder of this

chapter we will assume that we have one PIA available, having the RAM
addresses given below.

Address	*Register*
8009	Control register, CRA
800B	Control register, CRB
8008	$\begin{cases}\text{Data direction register, DDRA (CRA, } b_2 = 0) \\ \text{Input/output register, I/O A (CRA, } b_2 = 1)\end{cases}$
800A	$\begin{cases}\text{Data direction register, DDRB (CRB, } b_2 = 0) \\ \text{Input/output register, I/O B (CRB, } b_2 = 1)\end{cases}$

Let us now consider the function of each register.

b_7	b_6	b_5	b_4	b_3	b_2	b_1	b_0
IRQA(B) 1 Flag	IRQA(B) 2 Flag	CA2 (CB2) Control			DDR Access	CA1 (CB1) Control	

Figure 5.13 Control work format for the MC6821 PIA

The control register (CRA or CRB). The control word format shown in figure
5.13 may be summarised as follows. Bits b_0 and b_1 determine the CA1 (CB1)
operating mode. When b_0 = 0 this will disable the CA1 (CB1) interrupt line, and
in contrast when b_0 = 1 this will enable the CA1 (CB1) interrupt line. When
b_1 = 0 the *interrupt flag* (IRQA(B)1, bit b_7) is set by a high-to-low transition on
CA1 (CB1), and in contrast when b_1 = 1 the interrupt flag (IRQA(B)1, bit b_7) is
set by a low-to-high transition on CA1 (CB1).

Bit b_2 determines whether the data direction register or the I/O register is
addressed. If b_2 = 0 the data direction register is selected, however, if b_2 = 1 then
the I/O register is selected.

Bits b_3, b_4 and b_5 determine the CA2 (CB2) operating mode. When b_5 = 0
CA2 (CB2) is established as an input and b_3 and b_4 perform similarly to b_0 and
b_1. That is, if in this case b_3 = 0 the CA2 (CB2) interrupt line is disabled, but in
contrast, if in this case b_3 = 1, then the CA2 (CB2) interrupt line is enabled. Again
taking the case when b_5 = 0, then if b_4 = 0 the interrupt flag (IRQA(B)2, bit b_6)
is set by a high-to-low transition on CA2(CB2), and in contrast if in this case
b_4 = 1, then the interrupt flag (IRQA(B)2, bit b_6) is set by a low-to-high transition
on CA2 (CB2). Let us now consider the case when b_5 = 1. In this case CA2 (CB2)
acts as an output, and it functions in one of three modes. However, we will only
be concerned with the relatively simple set/reset mode. In this mode $b_5 = b_4 = 1$,
and b_3 serves as a program-controlled output, such that CA2 (CB2) is the set/reset
output which follows b_3 as it is changed by MPU *write to control register* operations.

Bit b_6 is the IRQA(B)2 interrupt flag which goes high on the active transition of CA2(CB2) if CA2(CB2) has been established as an input ($b_5 = 0$). This flag is reset to zero each time the MPU reads the corresponding I/O register, or it can be cleared by the hardware reset signal.

Bit b_7 is the IRQA(B)1 interrupt flag which goes high on the active transition of the enabled CA1(CB1) input. This flag is reset to zero each time the MPU reads the corresponding I/O register, or it can be cleared by the hardware reset signal.

The Data Direction Register (DDRA or DDRB). This register is used to estab-lish each peripheral bus line as either an input or an output. This is achieved by having the MPU write 0s or 1s into the 8 bit positions of the DDR. Thus if $b_j = 1$, where $j = 0, 1, \ldots, 7$, in the DDR, then b_j in the I/O register will be established as an output. In contrast if $b_j = 0$ in the DDR, then b_j in the I/O register will be established as an input.

The Input/Output Register (I/O A or I/O B). When this register is addressed the data present on the established PIA inputs may be loaded into an MPU accumula-tor via a LDA A(B) instruction. Alternatively, data may be transferred from an MPU accumulator to the established PIA outputs via a STA A(B) instruction.

It may seem that the PIA's programmability makes it a rather complex device to set up and operate. However, in the implementation of a digital filter only a small number of the large variety of available functions are utilised to successfully interface the MPU to the A/D and D/A converters.

The following example demonstrates how the M6802 microprocessor and associated LSI devices and peripheral circuits may be used to implement a simple digital filter.

Example 5.4

The filter has the following specification:

(i)   a low-pass characteristic with cut-off frequency, $f_{cd} = 100$ Hz,
(ii) sample period, $T = 1.6$ ms,
(iii) the design is to be based on a 2nd-order Butterworth low-pass prototype analogue filter; and
(iv) the bilinear Z-transform method of design is to be used.

SOLUTION
The transfer function of the denormalised 2nd-order Butterworth low-pass filter is

$$G(S) = \frac{\omega_{ca}^2}{S^2 + \sqrt{2}\,\omega_{ca}\,S + \omega_{ca}^2} \tag{5.12}$$

Because the bilinear Z-transform is to be used it is necessary to prewarp the analogue filter cut-off frequency, that is

$$\omega_{ca} = \frac{2}{T} \tan\left(\frac{\omega_{cd}T}{2}\right)$$

$$= \frac{2}{1.6 \times 10^{-3}} \tan\left(\frac{2\pi \times 100 \times 1.6 \times 10^{-3}}{2}\right)$$

$$= 687.2 \text{ rad/s}$$

The next step is to substitute this value of ω_{ca} in equation 5.12 to obtain the prewarped transfer function. This yields

$$G(S)_{pwt} = \frac{472243.84}{S^2 + 971.85\, S + 472243.84} \qquad (5.13)$$

For the bilinear Z-transform we use the substitution

$$S = \frac{2}{T} \times \frac{(Z-1)}{(Z+1)}$$

and for the specified value of T this gives

$$S = 1250 \times \frac{(Z-1)}{(Z+1)} \qquad (5.14)$$

Now substituting equation 5.14 in equation 5.13 we obtain

$$G(Z) = \frac{Z^2 + 2\,Z + 1}{6.88\, Z^2 - 4.62\, Z + 1.74} \qquad (5.15)$$

In this example the poles of $G(Z)$ are the roots of the denominator polynomial in equation 5.15, which are a complex conjugate pair at $Z = 0.336 \pm j0.374$. Thus it is seen that the poles lie within the unit-circle in the Z-plane, thereby yielding a stable filter.

The frequency response characteristic, $G(e^{j\omega T})$, is derived by substituting $e^{j\omega T}$ for Z in $G(Z)$. That is

$$G(e^{j\omega T}) = \frac{e^{j2\omega T} + 2\, e^{j\omega T} + 1}{6.88\, e^{j2\omega T} - 4.62\, e^{j\omega T} + 1.74}$$

The corresponding magnitude/frequency characteristic, evaluated for a set of frequency values in the range 0 to 400 Hz, is shown in figure 5.14.

Having established that the derived digital filter is stable and that it has the desired low-pass magnitude/frequency characteristic, we may now proceed with the derivation of the corresponding linear difference equation:

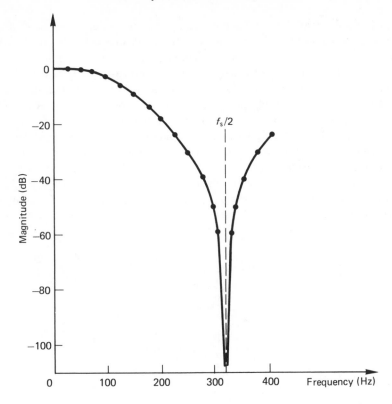

Figure 5.14 Magnitude/frequency characteristic for the digital filter in example 5.4

$$G(Z) = \frac{1 + 2\,Z^{-1} + Z^{-2}}{6.88 - 4.62\,Z^{-1} + 1.74\,Z^{-2}} = \frac{Y(Z)}{X(Z)}$$

Since Z^{-d} implies a time delay of d sampling periods it follows that

$$X(Z) + 2X(Z)Z^{-1} + X(Z)Z^{-2} \text{ transforms to}$$
$$x(n)T + 2x(n-1)T + x(n-2)T$$

and similarly

$$6.88\,Y(Z) - 4.62\,Y(Z)Z^{-1} + 1.74\,Y(Z)Z^{-2} \text{ transforms to}$$
$$6.88y(n)T - 4.62y(n-1)T + 1.74y(n-2)T$$

Therefore the linear difference equation of the filter is

$$y(n)T = 0.145x(n)T + 0.291x(n-1)T + 0.145x(n-2)T$$
$$+ 0.671y(n-1)T - 0.235y(n-2)T \tag{5.16}$$

We may now consider the implementation of the digital filter.

The block diagram of a suitable 6802-based microprocessor system is shown in figure 5.15. Basically, under program control the PIA generates the CA2-pulse (A/D convert command signal), thereby sampling the input signal to obtain the $x(n)T$ value. On completion of the A/D conversion process a status signal is generated by the A/D converter; this is a $0 \rightarrow 1$ transition which is accepted via the CA1 input to set the interrupt flag in the PIA control register. The resulting interrupt makes the program access the interrupt routine, which is used to read $x(n)T$ into the MPU via the designated PIA input lines (say PA0-PA7), and it is then used to evaluate $y(n)T$ (see equation 5.16), which is then output to the D/A converter via the designated PIA output lines (say PB0-PB7). The stored sampled-data values are updated and the microprocessor then returns from the interrupt routine. At the next sampling instant the process is repeated, as shown in the flowchart given in figure 5.16.

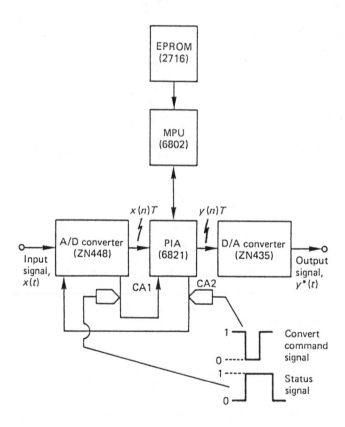

Figure 5.15 Block diagram of 6802-based digital filter system

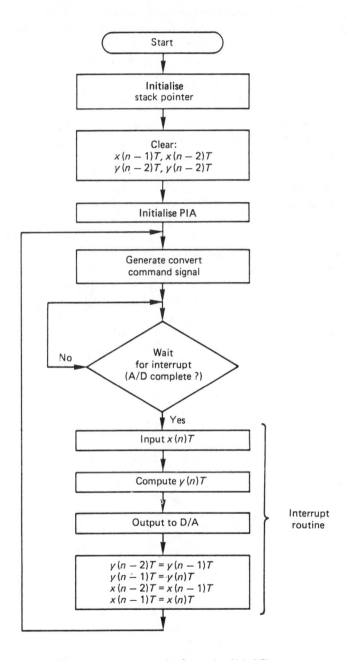

Figure 5.16 Flowchart for 2nd-order digital filter

Figure 5.17 shows relatively inexpensive hardware suitable for implementing the digital filter derived in this example. By inspecting figure 5.17 we may deduce that the memory-map of this system is

> 6802 RAM: 0000–007F (scratch-pad memory)
> 6821 PIA: 8008–800B (I/O interface)
> 2716 EPROM: F800–FFFF (program/constants memory)

Note that the system hardware shown in figure 5.17 does not include an anti-aliasing filter to band-limit the input signal. Consequently the highest frequency component in this signal must not exceed 312.5 Hz (see figure 5.14) if aliasing is to be avoided. Clearly this restriction is determined by the value of sampling frequency used, and the baseband frequency range of the input signal may be increased by a corresponding increase of the sampling frequency if the practical constraints of the system permit this higher rate of working. Also note that the system shown in figure 5.17 does not show the required reconstruction filter. This filter may be simply a 1st-order R–C low-pass circuit connected to the D/A converter output to smooth the stepped waveform produced at the D/A output. It is reasonable to assume that acceptable smoothing may be achieved when the reconstruction filter time-constant is approximately equal to ten times the value of the sampling period, that is $C \times R = 10 \times T$ seconds. It is appropriate to make R a variable resistor so that the time-constant may be adjusted to optimise the smoothing process. In this example if $C = 0.1\ \mu$F then a 10 kΩ fixed resistor connected in series with a 10 kΩ variable resistor permits the time-constant to be adjusted to achieve satisfactory smoothing.

With respect to computational efficiency it will be advantageous to implement equation 5.16 as

$$y(n)T \approx 0.140625x(n)T + 0.28125x(n-1)T + 0.140625x(n-2)T \atop + 0.671875y(n-1)T - 0.25y(n-2)T \quad (5.17)$$

The coefficient values in equations 5.17 have been chosen to aid the multiplication process by using shift-and-add operations to form the product. For example, $0.671875y(n-1)T$ may be evaluated as $(0.5 + 0.125 + 0.03125 + 0.015625)$ $y(n-1)T$, whereby if $y(n-1)T$ is a stored sampled-data value held in one of the two 6802 accumulators, this computation is achieved as

original stored $y(n-1)T$ value shifted one place right (= $0.5y(n-1)T$)
added to
original stored $y(n-1)T$ value shifted three places right (= $0.125y(n-1)T$)
added to
original stored $y(n-1)T$ value shifted five places right (= $0.03125y(n-1)T$)
added to
original stored $y(n-1)T$ value shifted six places right (= $0.015625y(n-1)T$)

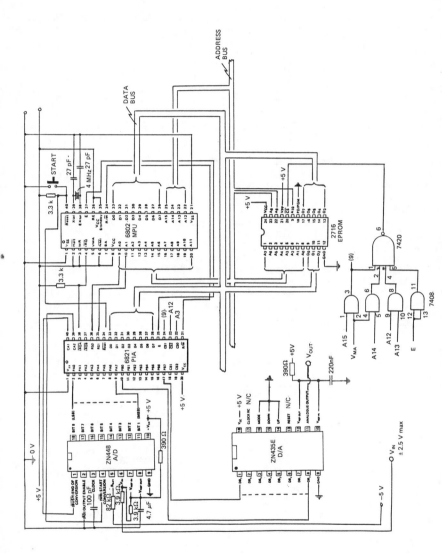

Figure 5.17   Hardware for an inexpensive microprocessor-based digital filter

```
        M68SAM IS THE PROPERTY OF MOTOROLA SPD, INC.
          COPYRIGHT 1974 AND 1975 BY MOTOROLA INC

        MOTOROLA M6800 CROSS ASSEMBLER, RELEASE 1.2

00001                       NAM     DIGFIL
00002      0000   X1        EQU     0          SCRATCH-PAD LOCATION (SPL)
00003      0001   X2        EQU     X1+1       SPL
00004      0002   Y1        EQU     X2+1       SPL
00005      0003   Y2        EQU     Y1+1       SPL
00006      0004   ACK       EQU     Y2+1       SPL
00007      0005   XN        EQU     ACK+1      SPL
00008      0060   SP        EQU     $60        STACK POINTER
00009      8008   PIA1      EQU     $8008      PIA I/OA OR DDRA
00010      8009   PIA2      EQU     $8009      PIA CRA
00011      800A   PIA3      EQU     $800A      PIA I/OB OR DDRB
00012      800B   PIA4      EQU     $800B      PIA CRB
00013 F800                  ORG     $F800
00014 F800 8E 0060 START    LDS     £SP        INITIALISE STACK POINTER
00015 F803 4F               CLR A
00016 F804 97 00            STA A   X1         (X1) = 0
00017 F806 97 01            STA A   X2         (X2) = 0
00018 F808 97 02            STA A   Y1         (Y1) = 0
00019 F80A 97 03            STA A   Y2         (Y2) = 0
00020 F80C 97 04            STA A   ACK        (ACK) = 0
00021 F80E B7 8009          STA A   PIA2       (CRA) = 0
00022 F811 B7 800B          STA A   PIA4       (CRB) = 0
00023 F814 B7 8008          STA A   PIA1       (DDRA) = 0,  A-SIDE AS INPUTS
00024 F817 43               COM A
00025 F818 B7 800A          STA A   PIA3       (DDRB) = FF, B-SIDE AS OUTPUT
00026 F81B 86 3C            LDA A   £$3C
00027 F81D B7 8009          STA A   PIA2       (CRA) = 3C   CA2 AS SET/RESET
00028              *                               OUTPUT = 1
00029              *                               CA1 DISABLED
00030              *                               I/OA SELECTED
00031 F820 B7 800B          STA A   PIA4       (CRB) = 3C   I/OB SELECTED
00032 F823 86 34   LOOP     LDA A   £$34
00033 F825 B7 8009          STA A   PIA2       CA2 OUTPUT = 0
00034 F828 86 3D            LDA A   £$3D
00035 F82A B7 8009          STA A   PIA2       CA2 OUTPUT = 1, CA1 ENABLED
00036              *                            CONVERT COMMAND
00037              *                            HAS BEEN GENERATED
00038 F82D 0E               CLI                CLEAR INTERRUPT MASK
00039 F82E 3E               WAI                WAIT FOR INTERRUPT
00040 F82F 0F               SEI                SET INTERRUPT MASK
00041 F830 7E F823          JMP     LOOP       REPEAT FOR NEXT SAMPLE
00042 F833 4F      COMP     CLR A              START SUBROUTINE
00043 F834 54               LSR B
00044 F835 54               LSR B
00045 F836 54               LSR B              (B) = 0.125 B
00046 F837 1B               ABA                (A) = 0.125 B
00047 F838 54               LSR B
00048 F839 54               LSR B
00049 F83A 54               LSR B              (B) = 0.015625 B
00050 F83B 1B               ABA                (A) = 0.140625 B
00051 F83C 39               RTS                RETURN FROM SUBROUTINE
00052 F83D F6 8008 INTR     LDA B   PIA1       START INTERRUPT ROUTINE,
00053              *                              (B) = XN
```

```
00054 F840 D7 05          STA B  XN       SAVE XN
00055 F842 BD F833         JSR    COMP
00056 F845 97 04          STA A  ACK      (ACK) = 0.140625 XN
00057 F847 25 43          BCS    SAT      IF OVERFLOW SATURATE OUTPUT
00058 F849 D6 00          LDA B  X1       (B) = X1
00059 F84B BD F833         JSR    COMP
00060 F84E 48             ASL A           (A) = 2(0.140625 X1)
00061 F84F 9B 04          ADD A  ACK      (A)=0.140625 XN + 0.28125 X1
00062 F851 97 04          STA A  ACK      SAVE (A) IN ACK
00063 F853 25 37          BCS    SAT      IF OVERFLOW SATURATE OUTPUT
00064 F855 D6 01          LDA B  X2       (B) = X2
00065 F857 BD F833         JSR    COMP
00066 F85A 9B 04          ADD A  ACK      (A)=0.140625 XN + 0.28125 X1
00067                 *                    + 0.140625 X2
00068 F85C 97 04          STA A  ACK      SAVE (A) IN ACK
00069 F85E 25 2C          BCS    SAT      IF OVERFLOW SATURATE OUTPUT
00070 F860 D6 02          LDA B  Y1       (B) = Y1
00071 F862 BD F833         JSR    COMP
00072 F865 9B 04          ADD A  ACK      (A)=0.140625 XN + 0.28125 X1
00073                 *                    + 0.140625 X2 + 0.140625 Y1
00074 F867 D6 02          LDA B  Y1       (B) = Y1
00075 F869 54             LSR B           (B) = 0.5 Y1
00076 F86A 1B             ABA             (A)=0.140625 XN + 0.28125 X1
00077                 *                    + 0.140625 X2 + 0.640625 Y1
00078 F86B 54             LSR B           (B) = 0.25 Y1
00079 F86C 54             LSR B           (B) = 0.125 Y1
00080 F86D 54             LSR B           (B) = 0.0625 Y1
00081 F86E 54             LSR B           (B) = 0.03125 Y1
00082 F86F 1B             ABA             (A)=0.140625 XN + 0.28125 X1
00083                 *                    + 0.140625 X2 + 0.671875 Y1
00084 F870 25 1A          BCS    SAT      IF OVERFLOW SATURATE OUTPUT
00085 F872 D6 03          LDA B  Y2       (B) = Y2
00086 F874 54             LSR B           (B) = 0.5 Y2
00087 F875 54             LSR B           (B) = 0.25 Y2
00088 F876 50             NEG B           (B) = -0.25 Y2
00089 F877 1B             ABA             (A)=0.140625 XN + 0.28125 X1
00090                 *                    + 0.140625 X2 + 0.671875 Y1
00091                 *                    - 0.25 = YN
00092 F878 2B 18          BMI    ZERO     IF MINUS ZERO THE OUTPUT
00093 F87A B7 800A        STA A  PIA3     OUTPUT YN
00094 F87D D6 02          LDA B  Y1       (B) = Y1
00095 F87F D7 03          STA B  Y2       (Y2) = Y1 UPDATE SAMPLE VALUE
00096 F881 97 02          STA A  Y1       (Y1) = YN UPDATE SAMPLE VALUE
00097 F883 D6 00          LDA B  X1       (B) = X1
00098 F885 D7 01          STA B  X2       (X2) = X1 UPDATE SAMPLE VALUE
00099 F887 D6 05          LDA B  XN       (B) = XN
00100 F889 D7 00          STA B  X1       (X1) = XN UPDATE SAMPLE VALUE
00101 F88B 3B             RTI             RETURN FROM INTERRUPT ROUTINE
00102 F88C 86 FF   SAT    LDA A  £$FF     (A) = FF
00103 F88E B7 800A        STA A  PIA3     SATURATE OUTPUT
00104 F891 3B             RTI             RETURN FROM INTERRUPT ROUTINE
00105 F892 4F     ZERO    CLR A           (A) = 0
00106 F893 B7 800A        STA A  PIA3     ZERO THE OUTPUT
00107 F896 3B             RTI             RETURN FROM INTERRUPT ROUTINE
00108                 *
00109 FFF8                ORG    $FFF8
00110 FFF8 F83D    IRQV   FDB    INTR     DEFINE INTERRUPT VECTOR
```

```
00111 FFFE                    ORG    $FFFE
00112 FFFE F800    RSTV       FDB    START    DEFINE RESET VECTOR
00113              *
00114                         OPT    LS       PRINT ASSM OUTPUT; SYMBOLS
00115              *
00116                         END
```

SYMBOL TABLE

X1	0000	X2	0001	Y1	0002	Y2	0003	ACK	0004
XN	0005	SP	0060	PIA1	8008	PIA2	8009	PIA3	800A
PIA4	800B	START	F800	LOOP	F823	COMP	F833	INTR	F83D
SAT	F88C	ZERO	F892	IRQV	FFF8	RSTV	FFFE		

Figure 5.18 Program for implementation of the digital filter in example 5.4

It is worth noting that four of the coefficients in equation 5.17 involve a value of 0.140625 (0.28125 = 2 × 0.140625; 0.671875 = 0.140625 + 0.53125), and it is clear that 0.140625 multiplied by a sampled-data value is a repetitive operation in the evaluation of $y(n)T$, consequently this is best achieved using a subroutine. The complete program listing for the implementation of the digital filter is given in figure 5.18, and it contains detailed comments to assist the reader in following the sequence of events outlined in the flowchart given in figure 5.16.

The performance of the digital low-pass filter may be demonstrated by using a square-wave input test signal having a period T_p, and defined by the following Fourier series:

$$x(t) = A_0 + \frac{2}{\pi}\left[\cos \omega_1 t - \frac{1}{3} \cos 3\omega_1 t + \frac{1}{5} \cos 5\omega_1 t - \ldots\right]$$

where $\omega_1 = 2\pi/T_p$. The considerable attentuation of the high-frequency nth odd harmonic components (of the form $1/n \cos n\omega_1 t$) in the input signal, resulting from the filtering process, produces an output signal $y(t)$ which is a reasonable approximation to a sinusoidal signal having a predominant fundamental component equal to the fundamental frequency of the input signal, that is

$$y(t) \approx A_0 + \frac{2}{\pi} \cos \omega_1 t$$

Figure 5.19 shows the output waveform $y(t)$ resulting from the implementation of the low-pass digital filter linear difference equation (equation 5.17), using the test signal $x(t)$ defined above.

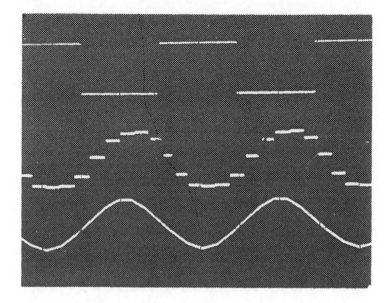

Figure 5.19 Waveforms for example 5.4. Top: input test signal, $x(t)$. Middle: D/A converter output signal, $y^*(t)$. Bottom: reconstruction filter output signal, $y(t)$

Another effective method of implementing a digital filter is to store all possible rounded product values in a ROM to form a *look-up table*. The appropriate product values are read from the ROM and their sum/difference is formed to directly yield the value of $y(n)T$. For example, suppose that a simple digital filter has the linear difference equation

$$y(n)T = 0.48x(n)T + 0.78y(n-1)T \qquad (5.18)$$

then it is possible to store product values, rounded to the nearest integer, as shown in the memory map, figure 5.20. Note that the notation $R(0.48(02)_{16})$ means the rounded value of (0.48 times the hexadecimal number 02), that is, the rounded value will be 01. Using the DIRECT addressing mode the value of $x(n)T$ can be used as the memory address to load accumulator A with its corresponding rounded product value at the appropriate point in the program. Also assuming that the microprocessor's index register has a pre-set value equal to $(0100)_{16}$, then using the indexed addressing mode the value of $y(n-1)T$ can be used as the second byte of the appropriate instruction, thereby loading accumulator B with its corresponding rounded product value at the appropriate point in the program. Accumulator B may be added to accumulator A to yield approximately (rounding errors are present) the value of $y(n)T$.

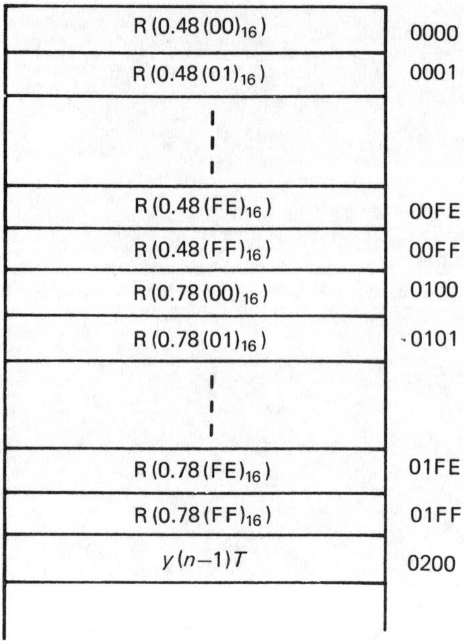

$R(0.48(00)_{16})$	0000
$R(0.48(01)_{16})$	0001
⋮	
$R(0.48(FE)_{16})$	00FE
$R(0.48(FF)_{16})$	00FF
$R(0.78(00)_{16})$	0100
$R(0.78(01)_{16})$	·0101
⋮	
$R(0.78(FE)_{16})$	01FE
$R(0.78(FF)_{16})$	01FF
$y(n-1)T$	0200

Figure 5.20 Memory map of rounded product values for equation 5.18

The program listing for the implementation of equation 5.18 is given in appendix 5.2, and the waveforms $x(t)$, $y^*(t)$ and $y(t)$ are shown in figure 5.21. This latter method of implementation obviously requires a considerable amount of memory to store the rounded product values. However, this does not pose any serious economic problem because the cost of suitable memory is not too restrictive, and as LSI technology continues to develop *packing density* is expected to increase, and costs are expected to be reduced further.

An added advantage of using a look-up table method of implementation is the relatively fast speed at which the value of $y(n)T$ can be obtained, and consequently higher sampling rates are possible.

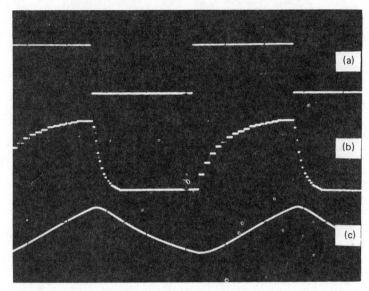

Figure 5.21 Waveforms for equation 5.18: (a) $x(t)$; (b) $y^*(t)$; (c) $y(t)$

5.4 DSP CHIPS

The first DSP chip was the Intel 2920[3], which was launched in 1979. It possessed on-chip data converters, but it lacked an on-chip multiplier and its instruction set was rather limited. The anticipated market for the 2920 was never achieved, but fortunately this did not deter other companies from developing their own DSP chips.

In 1980 NEC introduced their μPD7720[3,4] DSP chip. This incorporated a 16-bit $\times$ 16-bit multiplier and a 16-bit 2's complement ALU, and this combination permitted the implementation of a sum-of-products operation in a single instruction cycle (250 ns). At about the same time as the development of the NEC DSP chip, the implementation of an MOS integrated circuit for digital filtering was achieved at the British Telecom Research Laboratories. This chip, known by the acronym FAD[5,6] (filter and detect), requires a single-phase clock and very little external circuitry to implement a digital filter, see figure 5.22. The FAD chip is manufactured and marketed by Plessey Semiconductors as the MS2014FAD device. It is interesting to note that Plessey have launched other DSP products designed using their Megacell semicustom design methodology and manufactured on 2 μm CMOS. The first two members of the chip family are the two-dimensional edge detector (PDSP16401) and the 16 $\times$ 12 bit complex number multiplier (PDSP16112).

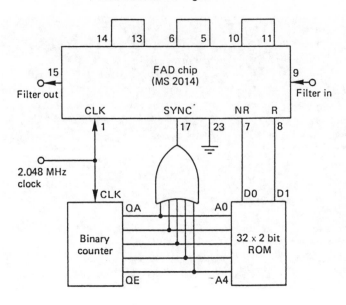

Figure 5.22 Block diagram of a typical FAD filter

Undoubtedly Texas Instruments has had remarkable success with its TMS320 family of DSP chips, which were initially launched in 1983 via the TMS32010. The TMS32010 is a microcomputer with a 32-bit internal Harvard architecture and a 16-bit external interface capable of executing five million instructions per second. The next significant development was the production of the TMS32020 DSP chip and, compared with the TMS32010, the TMS32020 has a number of improvements which makes it more suitable for DSP multiprocessing forms of implementation. Notably these permit synchronisation of the multiprocessing task, and include a serial port, DMA capability and a global data memory interface. The TMS320C25[7] is a pin-compatible CMOS version of the TMS32020 chip, which has a faster instruction cycle (100 ns), and also a on-chip memory comprising 544 words of data RAM, 4K words of masked ROM and 128K words of data/program space. The TMS320C25 is completely object-code-compatible with the TMS32020, and therefore any TMS32020 program will run on the TMS320C25. The instruction set of the TMS320C25 also supports adaptive filtering, extended-precision arithmetic, bit-reversed addressing and faster I/O for data-intensive signal processing.

5.4.1 TMS32010 Digital Signal Processor

To illustrate the features of a typical DSP device we will consider the popular and widely accepted TMS32010 chip. The other members of the TMS family are a natural extension to the TMS32010, and therefore a basic understanding of the

TMS32010 will equip the reader with knowledge that may be easily applied to other members of the family. Indeed, since many DSP devices have similar architectural features, then the basic knowledge gained in studying the TMS32010 is applicable to a wide range of DSP devices. This is analogous to the situation whereby a student gains knowledge of a particular microprocessor, and becomes familiar with it and its associated family of support chips, and then subsequently has to deal with a microprocessor produced by another manufacturer. In this case progression along the learning-curve is generally easier than starting from a zero base-level of knowledge because of the student's initial understanding of the basic principles of microprocessors.

Figure 5.23 shows the functional block diagram of the TMS320CM10 DSP device (the first low-power member of the family). It contains a 32-bit ALU

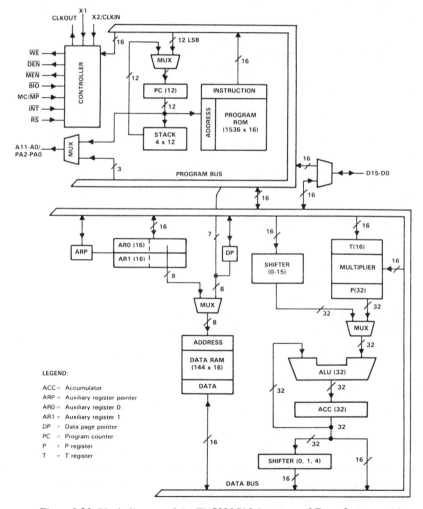

Figure 5.23 Block diagram of the TMS320C10 (courtesy of Texas Instruments)

and accumulator that support double-precision arithmetic. The ALU operates with the 16-bit words read from RAM or derived from immediate addressing instructions. The ALU performs two's complement fixed-point arithmetic and Boolean operations, thereby providing the number crunching and bit manipulation ability required in implementing DSP applications. The device contains two shifters. A barrel shifter is available for left-shifting data 0 to 15 places before it is loaded into, added to, or subtracted from the accumulator. This shifter extends the most-significant bit of the data word and zero-fills the least-significant bits for two's complement arithmetic. The second shifter left-shifts the upper (most-significant) half of the accumulator 0, 1 or 4 places while it is being stored in the data RAM.

The TMS320CM10 contains a hardware multiplier which performs a 16-bit × 16-bit two's complement multiplication in one 200 ns instruction cycle. The multiplicand is stored temporarily in the 16-bit T-Register; whereas the multiplier value is read from data memory or is derived from the multiply immediate instruction (MPYK); the 32-bit result is stored in the P-Register.

The TMS320CM10 can operate in one of two possible modes, with mode selection achieved via the MC/$\overline{\text{MP}}$ input pin. For the Microcomputer Mode (MC), instruction addresses 0–1535 correspond to the on-chip ROM and instruction addresses 1536–4095 correspond to off-chip memory. The 1536-word ROM is mask-programmed at the factory with a customer's program, which of course is only economic for large production runs, and therefore the other mode of operation is provided to facilitate full-speed operation from all 4096 off-chip instruction addresses. This alternative mode is known as the Microprocessor Mode ($\overline{\text{MP}}$), and it is suitable for prototyping and development work. The TMS320CM10 working in the $\overline{\text{MP}}$ mode is a ROMless version of the device and is referred to by the manufacturer as the TMS320C10. Instructions are stored as 16-bit words irrespective of the mode of operation. Similarly, data is held as 16-bit words in the 144-word on-chip RAM. Figure 5.24 shows the memory maps for each mode.

The chip supports three forms of memory addressing. These are:

(i)   Direct Addressing, which has the instruction format shown in figure 5.25a. The 144 words of RAM are subdivided into two pages (groups), and page selection is achieved via the 1-bit Data Page Pointer (DP); the first page contains 128 words, with the remaining 16 words in the second page. The DP Pointer may be set/reset via the LDP, LDPK or LST instructions.

(ii) Indirect Addressing, which has the instruction format shown in figure 5.25b. Again the 144 words of RAM are arranged in the two-page format, with selection via the DP Pointer. However, with this form of addressing one of the two Auxiliary Registers, AR0 or AR1, uses its least-significant eight bits to form the RAM address. Selection of the auxiliary registers is achieved using the 1-bit Auxiliary Register Pointer (ARP): ARP set to zero defines that AR0 holds the RAM address, whereas ARP set to one

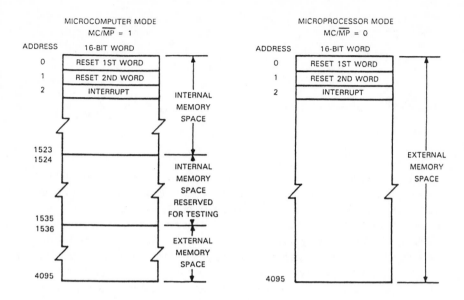

Figure 5.24 TMS320 family memory map (courtesy of Texas Instruments)

defines that AR1 holds the RAM address. Referring to figure 5.25b it should be noted that if bit 3 of the instruction-word is set to zero, then the content of bit 0 is loaded into the ARP, otherwise the ARP remains unchanged. Bits 4 and 5 of the instruction-word determine whether the selected auxiliary register is decremented by one or incremented by one; if bit 4 is set to one the selected auxiliary register is decremented, whereas if bit 5 is set to one the selected auxiliary register is incremented, however if bit 4 and bit 5 are both set to zero then the selected auxiliary register is neither incremented nor decremented. The incrementing/decrementing process is implemented in parallel (simultaneously) with the execution of any Indirect Addressing instruction, thereby permitting single-cycle manipulation of Data Tables.

(iii) Immediate Addressing, which has the data embedded in the instruction-word. For example, the LACK instruction would be used to load the accumulator with data using immediate addressing.

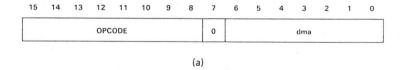

(a)

Figure 5.25(a) Direct addressing format (courtesy of Texas Instruments)

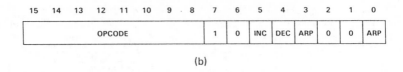

(b)

Figure 5.25(b)   Indirect addressing format (courtesy of Texas Instruments)

The device has a Status Register which consists of five status bits. The contents of this register may be stored in data memory using the store status instruction (SST), and the corresponding data word format is shown in figure 5.26. The Status Register may be reloaded or modified using the load status instruction (LST), except that the interrupt mask bit (INTM) can only be affected by the interrupt enable instruction (EINT) or the interrupt disable instruction (DINT). If the interrupt is enabled and an interrupt signal is received, then the INTM bit is set thereby disabling any subsequent interrupts until the bit is reset. In practice it is often appropriate to reset the INTM bit at the end of the Interrupt Service Routine. It is seen from figure 5.26 that the status register also contains the data page pointer and the auxiliary register pointer; the significance of their status was described above. Additionally the status register contains an Overflow Mode Bit (OVM) and an Overflow Flag Register (OV). The OVM bit is used to enable/disable the option of causing the accumulator to be loaded with the most positive/negative ALU value if an arithmetic overflow is detected. If the OVM bit is set equal to one this saturating option is enabled, otherwise it is disabled. The OV bit is set to one if the accumulator experiences an arithmetic overflow, otherwise the bit is reset to zero.

/ / / = don't care

Figure 5.26 Format of stored status word (courtesy of Texas Instruments)

The device has a 12-bit Program Counter (PC) which holds the address of program memory, and it is buffered to provide the A0-A11 address lines. Address lines A0-A2 are multiplexed with the input/output port address lines PA0-PA2. The device also contains a 4-level Stack for saving the program counter contents when interrupts or subroutine calls are serviced. Stack operations may also be achieved using data RAM via POP/PUSH and Store accumulator/Load accumulator instructions.

Figure 5.27 shows the N-package Top-View of the TMS320 DSP chip and its pin nomenclature. Some signals are discussed in more detail below.

The reset function is enabled when the $\overline{RS}$ pin is taken active-low for a minimum of five clock cycles. A reset causes outputs $\overline{DEN}$, $\overline{MEN}$ and $\overline{WE}$ to become active-high, and the data bus is tristated. Also the Program Counter (PC) and the address bus are cleared, consequently the processor then begins execution at location 0, see figure 5.24. Note that a branch instruction to the Reset Routine is usually placed in locations 0 and 1. A reset also disables the interrupt, clears the interrupt flag and leaves the overflow mode register unchanged. The reset state can be held indefinitely.

```
                    N PACKAGE
                    (TOP VIEW)

        A1/PA1  [ 1      40 ]  A2/PA2
        A0/PA0  [ 2      39 ]  A3
        MC/MP   [ 3      38 ]  A4
         RS     [ 4      37 ]  A5
         INT    [ 5      36 ]  A6
        CLKOUT  [ 6      35 ]  A7
          X1    [ 7      34 ]  A8
       X2/CLKIN [ 8      33 ]  MEN
         BIO    [ 9      32 ]  DEN
         Vss    [ 10     31 ]  WE
          D8    [ 11     30 ]  Vcc
          D9    [ 12     29 ]  A9
         D10    [ 13     28 ]  A10
         D11    [ 14     27 ]  A11
         D12    [ 15     26 ]  D0
         D13    [ 16     25 ]  D1
         D14    [ 17     24 ]  D2
         D15    [ 18     23 ]  D3
          D7    [ 19     22 ]  D4
          D6    [ 20     21 ]  D5
```

NAME	I/O	DEFINITION
A11-A0/PA2-PA0	O	External address bus. I/O port address multiplexed over PA2-PA0.
$\overline{BIO}$	I	External polling input for bit test and jump operations.
CLKOUT	O	System clock output, ¼ crystal/CLKIN frequency.
D15-D0	I/O	16-bit data bus.
$\overline{DEN}$	O	Data enable indicates the processor accepting input data on D15-D0.
$\overline{INT}$	I	Interrupt.
MC/$\overline{MP}$	I	Memory mode select pin. High selects microcomputer mode. Low selects microprocessor mode.
$\overline{MEN}$	O	Memory enable indicates that D15-D0 will accept external memory instruction.
NC		No connection.
$\overline{RS}$	I	Reset used to initialize the device.
V_{CC}	I	Power.
V_{SS}	I	Ground.
$\overline{WE}$	O	Write enable indicates valid data on D15-D0.
X1	I	Crystal input.
X2/CLKIN	I	Crystal input or external clock input.

Figure 5.27 TMS320C10 N-package and nomenclature (courtesy of Texas Instruments)

An interrupt is recognised when a negative-going edge is applied to the $\overline{\text{INT}}$ pin, or by holding the $\overline{\text{INT}}$ pin low, assuming that interrupts have been enabled by the INTM bit. When the Interrupt Service Routine begins, the INTM bit is automatically set, thereby disabling interrupts, and the processor interrupt flag is cleared. If the penultimate instruction in the interrupt service routine is EINT (enable interrupt), interrupt servicing will be delayed until the RET (Return) instruction is executed. An interrupt causes a branch to location 2 in program memory, see figure 5.24.

The $\overline{\text{BIO}}$ pin is a polling input which is sampled every clock cycle and is not latched. If this pin is in the low state when a BIOZ instruction is executed, then the program will branch to a specified memory address.

The device has a 200 ns instruction cycle time with a corresponding frequency of 5 MHz, which is derived from an external 20 MHz crystal, see figure 5.28. An external clock source can be used by injecting the signal directly into the X2/CLKIN pin with the X1 pin unconnected, but it must conform to specified timing requirements.[8] The device clock signal is available on the CLKOUT pin, and its frequency is one-fourth that of the crystal frequency value/external clock frequency.

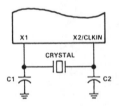

PARAMETER	TEST CONDITIONS	MIN	NOM	MAX	UNIT
Crystal frequency f_x	0 °C – 70 °C	6.7		20.5	MHz
C1, C2	0 °C – 70 °C			10	pF

Figure 5.28 TMS320 clock circuit (courtesy of Texas Instruments)

The input and output of data to and from peripheral devices are accomplished using the IN and OUT instructions. The $\overline{\text{MEN}}$ output signal is active-low on every machine cycle, but it is active-high when the $\overline{\text{WE}}$ and $\overline{\text{DEN}}$ signals are active. The $\overline{\text{DEN}}$ signal is active-low when the device is accepting data from the data bus as a result of executing an IN instruction, see figure 5.29a. The $\overline{\text{WE}}$ signal is active-low when the device has placed valid data on the data bus as a result of executing an OUT instruction, see figure 5.29b. The three multiplexed least-significant-bits of the address bus, PA0–PA2, are used as a port address by the IN and OUT instructions; the remaining address bus bits are held at logic

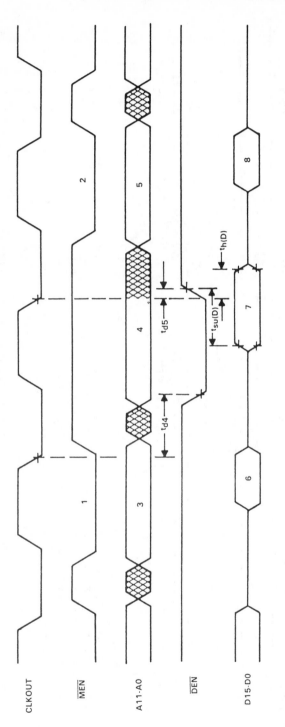

LEGEND:

1.	IN INSTRUCTION PREFETCH	5.	ADDRESS BUS VALID
2.	NEXT INSTRUCTION PREFETCH	6.	INSTRUCTION IN VALID
3.	ADDRESS BUS VALID	7.	DATA IN VALID
4.	PERIPHERAL ADDRESS VALID	8.	INSTRUCTION IN VALID

NOTE 2: Timing measurements are referenced to and from a low voltage of 0.8 volts and a high voltage of 2.0 volts, unless otherwise noted.

(a)

Figure 5.29(a) IN instruction timing signals (courtesy of Texas Instruments)

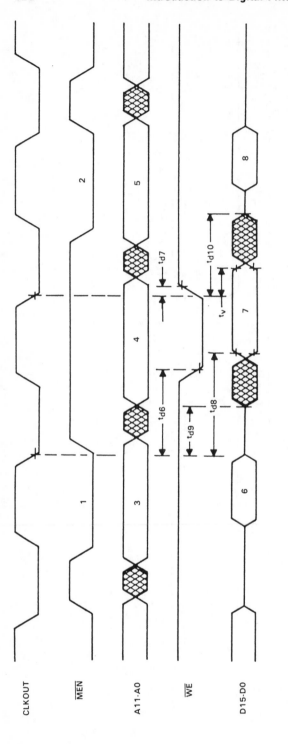

LEGEND:

1.   OUT INSTRUCTION PREFETCH
2.   NEXT INSTRUCTION PREFETCH
3.   ADDRESS BUS VALID
4.   PERIPHERAL ADDRESS VALID

5.   ADDRESS BUS VALID
6.   INSTRUCTION IN VALID
7.   DATA OUT VALID
8.   INSTRUCTION IN VALID

NOTE 2: Timing measurements are referenced to and from a low voltage of 0.8 volts and a high voltage of 2.0 volts, unless otherwise noted.

(b)

Figure 5.29(b)   OUT instruction timing signals (courtesy of Texas Instruments)

zero when either of these two instructions is executed. Data is transferred over the 16-bit data bus to one of the eight possible I/O ports, see figure 5.30.

The instruction set for the TMS320C10 DSP device is summarised in appendix 5.4, and it consists primarily of single-cycle single-word instructions, permitting execution rates in excess of 5 million instructions per second. Only infrequently used I/O and branch instructions are multicycle.

A practical example of the application of the TMS320C10 in telephone networks is illustrated in the following case study.

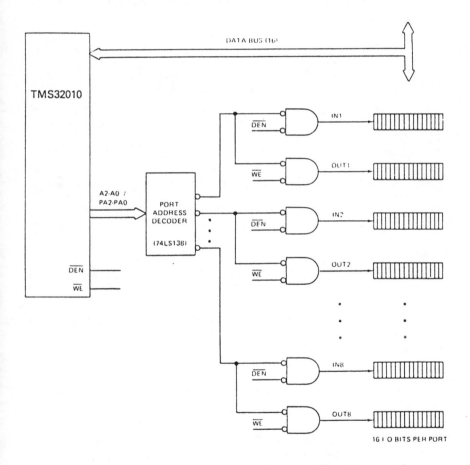

Figure 5.30 I/O interface for eight ports (courtesy of Texas Instruments)

5.4.1.1 TMS320C10 Case Study: Companding of Speech Signals

In Pulse Code Modulation (PCM) systems speech signals may be sampled and encoded as binary words for serial transmission over a telephone network,

commonly at a rate of 8000 samples per second. This encoded information is
most efficiently communicated when the sample amplitudes are compressed to a
logarithmic scale, resulting in a reduction in the number of bits required to
represent the sample amplitudes. Conversion to a logarithmic scale also ensures
that low amplitude speech signals are encoded with a minimal loss of fidelity. At
the receiver the encoded sample amplitudes are expanded to recover the trans-
mitted speech signal. This process of compressing and expanding the speech
signal is known as COMPANDING, the term being derived from COMpression
and exPANDING. Figure 5.31 shows the block diagram representation of the
PCM companding process.

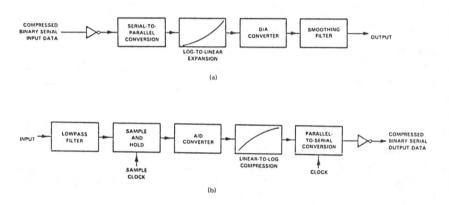

Figure 5.31 Block diagram of PCM companding system (courtesy of Texas Instruments)

Two International Standards for companding exist—in the USA and Japan it
is referred to as the μ-255 Law, whereas the European version is referred to as
the A-Law. The μ-255 Law is defined as

$$F(x) = \mathrm{sgn}(x) \, \frac{\ell n(1 + \mu \, |\, x \,|)}{\ell n(1 + \mu)} \qquad (5.19)$$

where $F(x)$ is the compressed output value, x is the normalised input value in the
range +1 to -1, μ is the compression parameter (equal to 255 in the USA), sgn(x)
is the sign (+ or $-$) of x, and ℓn denotes natural logarithm.

The A-Law is defined as

$$F(x) = \begin{cases} \text{sgn}(x) \ \dfrac{A \, |x|}{1 + \ell n(A)} & \text{for } 0 \leqslant |x| < 1/A \\[4mm] \text{sgn}(x) \ \dfrac{1 + \ell n \, A \, |x|}{1 + \ell n(A)} & \text{for } 1/A \leqslant |x| \leqslant 1 \end{cases} \qquad (5.20)$$

where $F(x), x$, $\text{sgn}(x)$ and ℓn are identical parameters to those used in the μ-Law definition. The value of A used in Europe is 87.6.

It will be instructive to consider a sample calculation for the A-Law companding process (equation 5.20). Suppose that $|x| = 0.0625$, and noting that $1/A \approx 0.01142$, then

$$F(0.0625) = \frac{1 + \ell n(87.6 \times 0.0625)}{1 + \ell n \, 87.6} \approx 0.49$$

Similar calculations may be undertaken for various values of $|x|$ to determine the form of the A-Law characteristic. This is commonly illustrated by the straight-line approximation shown in figure 5.32. Referring to this figure it is seen that the input sample value modulus, $|x|$, of successively larger intervals, is compressed into intervals of uniform size along the $F(x)$ axis.

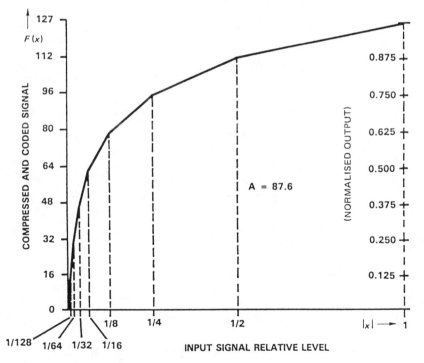

Figure 5.32 A-Law characteristic (courtesy of Texas Instruments)

Suppose that the A/D converter used in the PCM companding system produces a 13-bit representation of x in sign and magnitude format, see figure 5.33. Typically this 13-bit word is then compressed into a corresponding 8-bit representation in sign and magnitude format, thereby obtaining a processed input sample value in the form used by the transmitter to produce a compressed binary serial data output word. The required compression process in this case may be achieved using the method summarised in the flowchart given in figure 5.34.

SIGN BIT	BIT 11	BIT 10	BIT 9	BIT 8	BIT 7	BIT 6	BIT 5	BIT 4	BIT 3	BIT 2	BIT 1	BIT 0

Figure 5.33 A/D word format for sign and magnitude representation

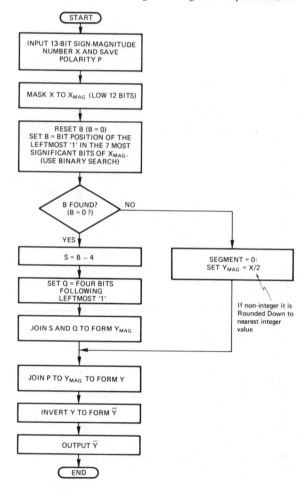

Figure 5.34 Flowchart for compression process

It will be instructive to consider the following two sample calculations, which use the compression method shown in figure 5.34. Suppose that X = 3121, then its binary representation in sign and magnitude format is manipulated as follows:

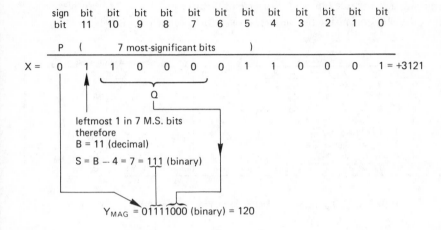

From the above we see that a sample value of 3121 is compressed to a value of 120.

Now suppose that X = −200, then using the same compression process we have

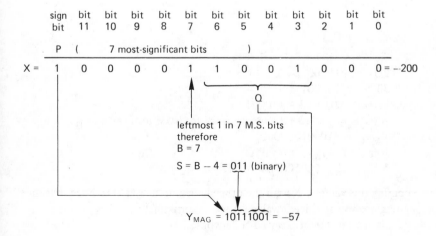

We see that a sample value of −200 is compressed to a value of −57.

From the above calculations we see that the 12-bit magnitude is compressed to a corresponding 7-bit magnitude, with the sign bit preserved. This compression process is summarised in figure 5.35 (the sign bit has been omitted).

Referring to figure 5.31 it is seen that the 8-bit code-word is inverted before transmission to increase the occurrence of positive pulses on the transmission line, thereby optimising the performance of the timing and recovery circuits.

Input Values												Compressed Code Word						
Bit: 11	10	9	8	7	6	5	4	3	2	1	0	Bit: 6	5	4	3	2	1	0
0	0	0	0	0	0	0	Q_3	Q_2	Q_1	Q_0	x	0	0	0	Q_3	Q_2	Q_1	Q_0
0	0	0	0	0	0	1	Q_3	Q_2	Q_1	Q_0	x	0	0	1	Q_3	Q_2	Q_1	Q_0
0	0	0	0	0	1	Q_3	Q_2	Q_1	Q_0	x	x	0	1	0	Q_3	Q_2	Q_1	Q_0
0	0	0	0	1	Q_3	Q_2	Q_1	Q_0	x	x	x	0	1	1	Q_3	Q_2	Q_1	Q_0
0	0	0	1	Q_3	Q_2	Q_1	Q_0	x	x	x	x	1	0	0	Q_3	Q_2	Q_1	Q_0
0	0	1	Q_3	Q_2	Q_1	Q_0	x	x	x	x	x	1	0	1	Q_3	Q_2	Q_1	Q_0
0	1	Q_3	Q_2	Q_1	Q_0	x	x	x	x	x	x	1	1	0	Q_3	Q_2	Q_1	Q_0
1	Q_3	Q_2	Q_1	Q_0	x	x	x	x	x	x	x	1	1	1	Q_3	Q_2	Q_1	Q_0

Figure 5.35 A-Law binary encoding table (courtesy of Texas Instruments)

The received 8-bit code-word is expanded into a corresponding 13-bit representation, in sign and magnitude format, using the process summarised in the flowchart given in figure 5.36. The following sample calculations illustrate the A-Law expansion process in this case.

Suppose that the received code-word is

$\overline{Y}$ = .10000111 (corresponds to compression of 3121)

then this is processed as follows:

Y   = 01111000

P   = 0

S   = 111 = 7; S − 1 = 6

2Q   = 10000

33   = 100001

SUM   = 2Q + 33 = 110001

ASL (S − 1) places the binary word for SUM to form X_{MAG}:

X_{MAG} = 110001000000

Join P to X_{MAG} to form X:

X   = 0110001000000 = 3136

By comparing this value of X with the original magnitude (3121) we see that the companding process produces a value error of approximately 0.5 per cent.

Now consider a different received code-word:

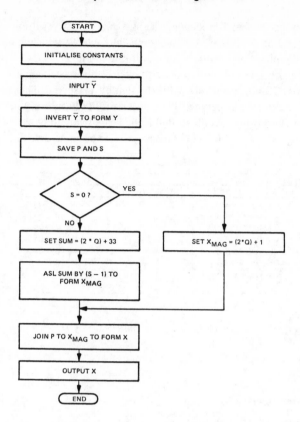

Figure 5.36 Flowchart for expansion process

$$\overline{Y} = 01000110 \text{ (corresponds to compression of } -200)$$

this is processed as follows:

$$Y = \underline{1011\,1001}$$

$$\begin{array}{c} \downarrow \quad \big| \quad Q \end{array}$$

$$P = 1 \downarrow$$

$$S = \overbrace{011} = 3; S - 1 = 2$$

$$2Q = 10010$$

$$\text{SUM} = 2Q + 33 = 110011$$

$$\text{ASL } (S - 1) \text{ places SUM} = 11001100$$

$$X_{MAG} = 000011001100$$

$$X = 1000011001100 = -204$$

By comparing this value of X with the original magnitude (200) we see that the companding process produces a value error of approximately 1.96 per cent.

However, in speech signal processing the loss of accuracy resulting from the companding operation is often acceptable, especially when the saving in the number of bits for sample value representation is considerable.

From the above calculations we see that the 7-bit magnitude is expanded to a corresponding 12-bit magnitude, with the sign bit preserved. This expansion process is summarised in figure 5.37 (the sign bit has been omitted).

The companding process illustrated in this case study may be readily implemented using the TMS320C10 DSP device, see the program given in appendix 5.5.

Compressed Code Word							Output Values											
Bit: 6	5	4	3	2	1	0	Bit: 11	10	9	8	7	6	5	4	3	2	1	0
0	0	0	Q_3	Q_2	Q_1	Q_0	0	0	0	0	0	0	0	Q_3	Q_2	Q_1	Q_0	1
0	0	1	Q_3	Q_2	Q_1	Q_0	0	0	0	0	0	0	1	Q_3	Q_2	Q_1	Q_0	1
0	1	0	Q_3	Q_2	Q_1	Q_0	0	0	0	0	0	1	Q_3	Q_2	Q_1	Q_0	1	0
0	1	1	Q_3	Q_2	Q_1	Q_0	0	0	0	0	1	Q_3	Q_2	Q_1	Q_0	1	0	0
1	0	0	Q_3	Q_2	Q_1	Q_0	0	0	0	1	Q_3	Q_2	Q_1	Q_0	1	0	0	0
1	0	1	Q_3	Q_2	Q_1	Q_0	0	0	1	Q_3	Q_2	Q_1	Q_0	1	0	0	0	0
1	1	0	Q_3	Q_2	Q_1	Q_0	0	1	Q_3	Q_2	Q_1	Q_0	1	0	0	0	0	0
1	1	1	Q_3	Q_2	Q_1	Q_0	1	Q_3	Q_2	Q_1	Q_0	1	0	0	0	0	0	0

Figure 5.37 A-Law binary decoding table (courtesy of Texas Instruments)

5.4.2 Future Trends

The emergence of VLSI microelectronics technology and computer-aided-design (CAD) methodologies has given rise to significant developments of DSP and VLSI signal processing. However, it would be foolish to expect all further advances to emerge solely from improvements in device fabrication techniques. Consequently it is expected that major significant advances will be achieved from the development of novel architectural structures and the efficient implementation of parallelism in the computational processes. Designers have necessarily had to produce improved processor architectures, and notably the classical sequential processing strategy (von Neumann form) has been replaced by the more efficient parallel processing form (Harvard architecture), thereby increasing the speed of data handling and computations. For example, the Motorola DSP56000 device[9] architecture has been designed to maximise throughput in data-intensive DSP applications. This objective has been realised by the provision of two independent expandable data memory spaces, two address arithmetic units, and a data ALU which has two accumulators and two shifter/limiters. This duality of the device architecture facilitates efficient writing and execution of software for DSP applications. Research has shown that parallel processing architectures with highly attractive and promising characteristics, for utilising VLSI technology, may arise from a combination of systolic and wavefront architectures.[10,11]

A systolic system may be configured using an array of processors operating in a rhythmical manner to perform the computations and transmission of data for the desired signal processing operation. Basically all of the processing elements use common control (timing) signals and simultaneously perform the same function on different sampled-data values. The systolic array therefore exhibits modularity, regularity, local interconnection synchronised pipelined multi-processing, and offers remarkable potential for improved architectures for future DSP chips and VLSI signal processing.[12-14]

Manufacturers and suppliers of DSP devices are an excellent source of information and produce advance data pertaining to new developments. A list of useful names, addresses and telephone numbers is given in appendix 5.6 (details may change with time, but it should be possible to contact manufacturers by referring to the Trade Press for addresses or by checking telephone numbers with the Telephone Directory Service).

5.5 SIGNALS AND NOISE

A digital filter may be used to eliminate or reduce unwanted *noise* which may be contaminating a wanted signal. The noise signal may appear in a variety of forms, for example, it may occur as mains-frequency hum, which in the United Kingdom will have a predominant 50 Hz frequency component; alternatively the noise could have been produced by random variations in electrode contact potential in a biomedical instrumentation system, typically creating noise in the range 10 to 500 Hz.

One of the most difficult problems associated with trying to eliminate noise using a digital filter is the process of choosing the appropriate filter specification. The difficulty generally arises from insufficient knowledge of the signal power spectrum and the noise power spectrum, and often an estimate of their relative levels and frequency range may be no more than a reasonable guess. For example, in a particular case it may be that we can intuitively guess that the noise signal is randomly occurring having a typical rms value somewhere in the region of 10 dB with a spectrum which is approximately flat from 0 to 600 Hz, and furthermore we may be able to guess that the wanted signal is approximately 30 dB above the noise level, containing no frequencies of interest above, say, 12 Hz. In any case if the filter implementation turns out to be inadequate then it may be that improved performance can only be achieved on an experimental basis.

To enhance the signal waveform in the presence of noise using a digital filter it is desirable that the wanted signal and unwanted noise ideally occupy separate frequency ranges. This means that the power spectra for the signal and noise should be separated. An example of a situation that approaches closely to the ideal is the measurement of the stroboradiocardiogram[15] (a medical electronics application) where the signal and noise power spectra are separated into two distinct sections as shown in figure 5.38. Referring to figure 5.38 we see that a lowpass digital filter having a cutoff frequency of approximately 5 Hz will eliminate the unwanted noise. However, in many cases the noise power spectrum significantly overlaps the signal power spectrum, and therefore the noise will only be partially reduced by the filtering action (see figure 5.39).

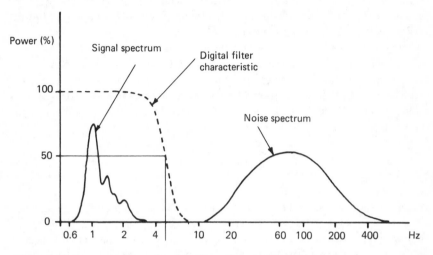

Figure 5.38 Power spectrum of stroboradiocardiogram signal and noise

If it is known that the wanted signal, which is contaminated by noise, repeats periodically, then it may be possible to use a signal averaging technique[16] to assist the *signal recovery* process. At this point it is appropriate to mention that the term 'signal recovery' generally refers to the process of trying accurately to extract an unknown signal waveform from contaminating noise. In contrast, the term *detection* is normally associated with the process of trying accurately to determine the time of occurrence of a known periodic time-limited signal waveform contaminated by noise. The *matched digital filter* may sometimes be used to good effect in a signal detection application, as discussed below.

5.5.1 The Matched Digital Filter

Suppose that a known time-limited signal waveform $x(t)$ is applied to the input of a single-input single-output linear system which has an impulse response, $g(t)$, equal to a time-reversed replica of $x(t)$, that is, we are considering the case where $g(t) = x(-t)$; this type of system is referred to as a *matched filter*. The output of the matched filter may be obtained by applying the well known convolution integral; however, in this particular case this procedure is identical to that used in obtaining the autocorrelation function, $R_{xx}(\tau)$ of $x(t)$. The maximum value of an autocorrelation function occurs when $\tau = 0$, that is $R_{xx}(\tau)_{max} = R_{xx}(0)$. This

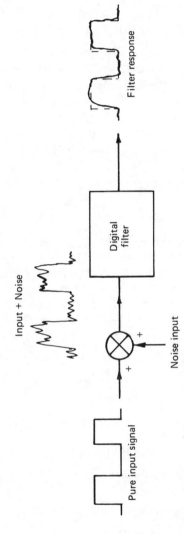

Figure 5.39 Partial noise reduction

means that the maximum output of the matched filter corresponds to $R_{xx}(0)$, and if the noise has a constant power spectral density over the signal frequency range, then the improvement in the signal-to-noise ratio will be optimum.

The sampled-data input to the matched digital filter consisting of N samples, may be denoted as the set of values

$$\{x(1)T, x(2)T, x(3)T, \ldots, x(N)T\}$$

with the maximum value in the set being denoted as $x(M)T$. We may assume that the input noise samples are uncorrelated with a standard deviation σ. The ratio of the maximum input sample to standard deviation of the input noise is

$$\frac{x(M)T}{\sigma} \tag{5.21}$$

The matched digital filter will have an impulse response consisting of the set of values

$$\{x(N)T, x(N-1)T, x(N-2)T, \ldots, x(1)T\}$$

and thus the maximum output of the filter is given by

$$R_{xx}(0) = [x(1)T^2 + x(2)T^2 + x(3)T^2 + \ldots + x(N)T^2] \tag{5.22}$$

The filter will produce output noise samples, each one being equal to the sum of N weighted input noise samples—the impulse response terms being the applied weighting coefficients. Now recalling that each input noise sample is assumed to have a standard deviation σ, then it follows that each input noise sample will have a variance equal to σ^2, and since variances are additive, then the filter's output noise samples are equal to

$$[\sigma x(1)T]^2 + [\sigma x(2)T]^2 + \ldots + [\sigma x(N)T]^2$$
$$= \sigma^2 [x(1)T^2 + x(2)T^2 + \ldots + x(N)T^2]$$

and the corresponding standard deviation of noise at the filter output is

$$\sigma [x(1)T^2 + x(2)T^2 + \ldots + x(N)T^2]^{1/2} \tag{5.23}$$

From equations 5.22 and 5.23 we see that the ratio of maximum signal output to noise standard deviation at the output is

$$\frac{[x(1)T^2 + x(2)T^2 + \ldots + x(N)T^2]}{\sigma[x(1)T^2 + x(2)T^2 + \ldots + x(N)T^2]^{1/2}} = \frac{[x(1)T^2 + x(2)T^2 + \ldots + x(N)T^2]^{1/2}}{\sigma} \tag{5.24}$$

From equations 5.21 and 5.24 we may deduce that the matched digital filter has improved the signal-to-noise ratio by a factor of

$$\frac{[x(1)T^2 + x(2)T^2 + \ldots + x(N)T^2]^{1/2}/\sigma}{x(M)T/\sigma}$$

$$= \frac{[x(1)T^2 + x(2)T^2 + \ldots + x(N)T^2]^{1/2}}{x(M)T} \tag{5.25}$$

Hence we see from equation 5.25 that the signal-to-noise ratio depends only on: (1) the energy in the signal waveform; and (2) the maximum input sampled-data value. Furthermore, it can readily be seen that the benefit derived by using a matched digital filter does not depend on the shape of the input signal, for example, compare the improvement in the signal-to-noise ratio achieved by the filter with sampled-data inputs of $\{-1, -2, +3, -2, -1, +1\}$ or $\{-3, -2, -2, 0, 1, 2, -1\}$ –theoretically both would experience the same improvement.

Example 5.5 below, demonstrates a typical practical application of the matched digital filter.

Example 5.5
A pure electrocardiogram (ECG) waveform may typically take the form shown in figure 5.40. Suppose that it is contaminated by noise, and that it is required to detect the occurrence of the ECG waveform in the noise and thereby determine the time period between successive peak values, R. The assumed data for this problem is

$x(n)T = \{0.2, 0.2, 5.5, -0.2, 0.6, 1.2, 0, 0, 0, 0,\}$ repeats

$x(n)T$ + noise = $\{0.4, -0.2, 3.2, 0.7, 1.6, 2.9, 0.2, -0.3, -0.5, 0.2, 0.8, 1.4,$

$$2.6, 0.1, -0.4, 2.4, 0.5, 0.9, \ldots\}$$

Show how a matched digital filter may be used to achieve the desired result. Comment on the method used.

SOLUTION
From the sampled-data values of the ECG waveform we know that the impulse response of the filter will be a time-reversed replica of it, therefore the corresponding pulse transfer function of the matched digital filter will be

$$G(z) = 1.2Z^{-1} + 0.6Z^{-2} - 0.2Z^{-3} + 5.5Z^{-4} + 0.2Z^{-5} + 0.2Z^{-6}$$

and the corresponding linear difference equation for this filter is

$$y(n)T = [1.2x(n-1)T + 0.6x(n-2)T - 0.2x(n-3)T + 5.5x(n-4)T$$
$$+ 0.2x(n-5)T + 0.2x(n-6)T] \tag{5.26}$$

Using equation 5.26 the filter output values, $y(n)T$, may be calculated, thus giving the response summarised in table 5.7. Now referring to table 5.7 we see that it is possible to specify a *threshold level* whereby only output values $[y(n)T]$ exceed-

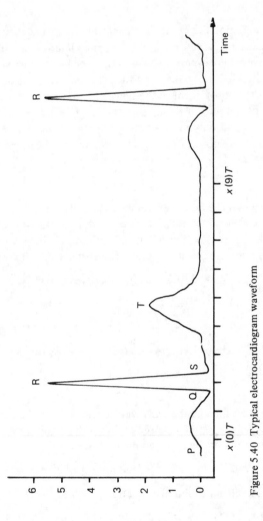

Figure 5.40 Typical electrocardiogram waveform

ing this level are taken to indicate the occurrence of point R in the ECG waveform. In this example a suitable threshold level may be chosen to be 16.5, and hence only $y(6)T$ and $y(16)T$ in table 5.7 will exceed this value, and thus the corresponding time interval between the two successive peak ECG values, R, is 10 times the sampling period, and the sampling period is usually a pre-determined value (or easily measured), consequently the desired time period between two successive R peaks can be determined.

Table 5.7 Response of matched digital filter in example 5.5

$$y(0)T = 0$$
$$y(1)T = 0.48$$
$$y(2)T = 0$$
$$y(3)T = 3.64$$
$$y(4)T = 5$$
$$y(5)T = 0.68$$
$$y(6)T = 21.94$$
$$y(7)T = 6.11$$
$$y(8)T = 8.76$$
$$y(9)T = 15.59$$
$$y(10)T = 2.12$$
$$y(11)T = 0.15$$
$$y(12)T = -0.65$$
$$y(13)T = 4.59$$
$$y(14)T = 7.15$$
$$y(15)T = 6.96$$
$$y(16)T = 17.36$$
$$y(17)T = 3.47$$
$$y(18)T = -0.76$$
$$\vdots$$

Obviously the choice of threshold level is an important factor in this application, and clearly there is a statistical chance that the noise will be worse than anticipated and consequently an error in judgement may exist. However, if possible this error should be reduced by attempting to minimise the noise in the input signal, perhaps by an initial filtering action.

5.6 CONCLUDING REMARKS

In this chapter the hardware and software aspects of digital filter implementations have been introduced. The main differences between dedicated hardware implementation and microprocessor implementation have been discussed and demonstrated. Also it has been shown that the implementation of a linear difference equation is readily achieved using a microprocessor, for example the implementation of the matched digital filter (represented by equation 5.26) is easily achieved using this method.

However, it is also appropriate at this point to mention that an alternative form of implementation exists, which uses a charge-coupled device (CCD) as the basic building block for the filter, it should also be realised that digital filters and CCDs are both allied to the subject of sampled-data signal processing, and consequently it is generally appropriate to consider both forms when trying to implement a particular filtering operation. In evaluating the CCD it can be helpful to have a basic knowledge of CCD operation; however, this topic is outside the scope of this book and therefore the interested reader is advised to consult the article by Burt.[17]

Another notable point is that it is claimed that the main advantage of using CCDs in sampled-data signal processing applications is reduced cost compared with an equivalent digital implementation. However, CCDs do have performance limitations when compared with digital filters, and therefore, because of their limited applicability, it is not expected that CCDs will make digital filters obsolete. This is especially true when we consider what the microprocessor now offers, and what it is likely to offer in the future. Indeed, the inherent advantages (programmability, compactness, low cost, etc.) of a microprocessor based digital filter system will make this an attractive method of implementation in the future.

REFERENCES

1. D. Queyssac (ed.), *Understanding Microprocessors* (Motorola Inc., 1976), chapter 2.
2. R. J. Simpson and T. J. Terrell, *Introduction to 6800/6802 Microprocessor Systems* (Newnes Technical Books – Butterworth & Co, Sevenoaks, 1982).
3. D. Quarmby, *Signal Processor Chips* (Granada Publishing, London, 1984).
4. R. J. Simpson and T. J. Terrell, 'Digital Filtering using the NEC μPD7720 signal processor', *Microprocessing and Microprogramming*, 14 (1984) pp. 67–78.
5. 'Digital Signal Processing with the MS2014 FAD', *Electronic Product Design*, (May 1986) pp. 49–56.
6. P. F. Adams, J. R. Harbridge and R. H. Macmillan, 'An MOS Integrated Circuit for Digital Filtering and Level Detection', *IEEE Journal of Solid-State Circuits*, SC-16, 3, (1981) pp. 183–190.
7. *TMS320C25 Digital Signal Processor – Product Description*, Texas Instruments, Bedford (1986).
8. *TMS32010 User's Guide*, Texas Instruments, Bedford (1983).
9. *DSP56000 Product Description*, Motorola, Glasgow (1986).
10. H. T. Kung, 'Systolic Arrays for VLSI', *Sparse Matrix Proc: SIAM*, Philadelphia (1979).
11. S. Y. Kung *et al.*, ' 'Wavefront Array' Processor: Language, Architecture and Applications', *IEEE Trans. Comput.* (1982) pp. 1054–66.
12. S. Y. Kung, H. J. Whitehouse and T. Kailath, *VLSI and Modern Signal Processing*, (Prentice-Hall, Englewood Cliffs, N.J., 1985).
13. E. E. Swartzlander, Jr, *VLSI Signal Processing Systems* (Kluwer Academic Publishers, Publishers, Hingham, Massachusetts, 1986).
14. P. Denyer and D. Renshaw, *VLSI Signal Processing: a bit-serial approach* (Addison-Wesley, Wokingham, 1985).
15. M. Della Corte and O. Cerofolini, 'Application of a Digital Filter to Biomedical Signals', *Med. biol. Engng.* (1974) 374–7.

16. P. A. Lynn, *An Introduction to the Analysis and Processing of Signals* (Macmillan, London, 1973) chapter 10.
17. D. J. Burt, 'Basic Operation of the Charge Coupled Device', *Conference Publication: Technology and Applications of Charge Coupled Devices*, University of Edinburgh, (1974) 1–12.

PROBLEMS

5.1 Referring to example 5.1, what word length is required to remove the error in the calculation of the $y(3)T$ value?

5.2 Write an M6802 subroutine program to perform the arithmetic: $A = 12(19B - 2C)$.

5.3 Write an M6821 PIA initialisation program such that I/O lines (PA0–PA4) and (PB6–PB7) are inputs, and the I/O lines (PA5–PA7) and (PB0–PB5) are outputs. Also initialise the PIA such that CA2 and CB2 are set/reset flip-flops, CA1 is inhibited and CB1 is an interrupt input which responds to a $0 \rightarrow 1$ transition.

5.4 The linear difference equation of a simple digital filter is

$$y(n)T = 0.39x(n)T + 0.88y(n - 1)T - 0.45y(n - 2)T$$

(a) Draw a flowchart which can be used as the basis for implementing the filter using the M6802 microprocessor system referred to in example 5.4; and (b) develop an M6802 program to implement the filter.

5.5 A sampled-data signal represented by the set

$$\left\{3, 1, 0, -1, -2\right\}$$

is to be detected when it is contaminated by wideband noise. Obtain the linear difference equation of a matched digital filter to achieve detection of this signal. What factor of improvement in the ratio of maximum input sample value to standard deviation of the noise will be produced by this filter?

APPENDIX 5.1 A BRIEF OVERVIEW OF SOME HARDWARE COMPONENTS

Set/Reset Latch

A simple form of set/reset *latch* is the SR *flip-flop* (shown in figure 5.41) which is used to store a single bit of data. The truth table, which summarises the operation of the flip-flop, is shown in figure 5.41, and it should be noted that Q^+ is the value of output Q after the application of an input change on S or R. The condition $S = R = 0$ produces an indefinable output state (denoted by $*$ in the truth table), and this must be avoided in logic design.

The latch may be reset ($Q = 0$) by applying a $1 \rightarrow 0 \rightarrow 1$ pulse to R, and subsequently the latch can be set ($Q = 1$) by applying a $1 \rightarrow 0 \rightarrow 1$ pulse to S (see figure 5.41).

The latch will memorise (store) the occurrence of the set pulse, and will remain set until reset. It is important to note that the latch is *volatile*, which means that stored data is lost when the power supply to the NAND gates is removed.

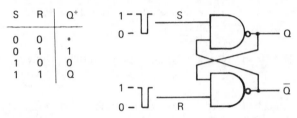

S	R	Q^+
0	0	*
0	1	1
1	0	0
1	1	Q

Figure 5.41 Set/reset latch

In practice a number of SR flip-flops can be contained on a single 'chip', for example, the SN74118 is a hex SR latch.

Shift Registers

A *shift register* consists of volatile *bistable* (flip–flop) stores which may be set to contain a binary word, and by application of subsequent clock pulses the word can be shifted successively from a bistable to the following one. The arrangement of a *shift-right-register* using JK flip-flops is shown in figure 5.42a. Each bistable has a set and reset input and hence initial conditions can be loaded into the register. A JK flip–flop is triggered on the trailing edge of the clock pulse, and the binary word will shift one place right with each applied pulse.

Shift registers can also be implemented using D-type flip–flops (see figure 5.42b, but in this case shifting of the stored binary word takes place on the leading edge of the clock pulse.

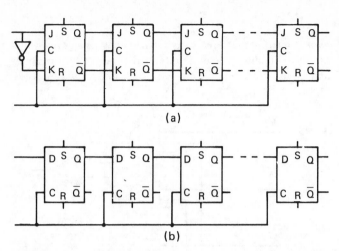

Figure 5.42 (a) Shift-register implemented using JK flip-flops. (b) Shift-register implemented using D-type flip-flops

In digital filter implementations a parallel-load 8-bit shift register may be appropriate, and typically this could be an SN74165 (MSI) TTL circuit.

Data Selector

A *data selector*, sometimes referred to as a *multiplexer*, is a device which is used to route data by switching selected inputs to the output. For example, the SN74157 may be used to select a 4-bit word from one of two sources and route it to the output, see figure 5.43.

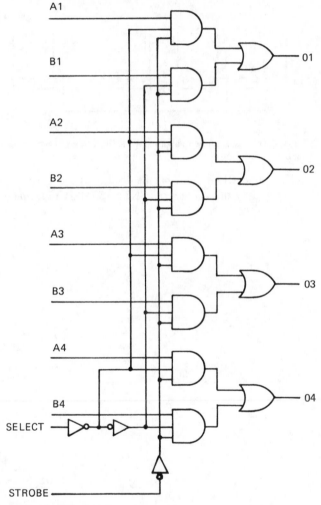

Figure 5.43 Quadruple 2-line-in-1-line selector

Read Only Memory (ROM)

A bipolar read only memory is a semiconductor integrated circuit which has a number of *cells* arranged as a 2^I (w + 1)-bit matrix configuration; each cell is capable of being set to binary 1 or binary 0. In a microprocessor system the stored

1s or 0s in the ROM form the program of instructions, and this program may be either permanent or alterable, depending upon the type of ROM used. Three types of ROM are commonly used as follows.

(1) *Mask-programmed ROMs* are programmed at the manufacturing stage, the final bit pattern being determined by the layer of metallisation produced using a custom designed mask. Each separate program requires its own unique mask, and consequently this type of ROM fabrication is costly, and is only used when large quantities of identical ROM stored programs are used. A typical ROM of this type is the MCM68308, 1024 × 8-bit store. One advantage of this ROM is that it is non-volatile so that the contents of memory are retained when the power supply is removed. However, in contrast, the main disadvantage is that the stored information cannot be altered, and obviously in this case there can be no allowance for error in the programming phase of system development.

(2) *Programmable ROMs (PROMs)* are programmed by the user by selectively fusing (burning out) the polysilicon or metal links in each memory cell, thereby permanently setting the fused cell to the desired state. A typical ROM of this type is the SN74186, 64 × 8-bit store. A PROM programmer instrument is required to achieve the fusing operation, but its purchase price will be relatively small when compared with mask development costs.

(3) *Erasable programmable ROMs (EPROMs)* store information by building up electric charge in a MOSFET cell. This stored charge can be erased by subjecting the cell to an intense source of ultraviolet radiation (PROM Eraser Instrument) for approximately 25 min. After a program has been erased a new program may be stored using a PROM programmer instrument. Using an EPROM in prototype development has the advantage of allowing program changes to be made, thus avoiding unnecessary 'chip' wastage. A typical ROM of this type is the MCM2716, 2048 × 8-bit store.

Random Access Memory

A random access memory may be either a *static* or *dynamic* form. The dynamic RAM has information stored as electric charge on the gate capacitance of MOS transistors. However, this capacitance will leak charge, and hence it is necessary periodically to refresh the memory cells to prevent loss of stored information. The static form of RAM uses conventional semiconductor bistable memory cells which do not require refreshing.

Normally a dynamic RAM consumes less power than its static counterpart, but it does require comparatively more associated circuitry which is used by synchronising the memory refresh cycles with the MPU read/write operations.

Semiconductor RAM is volatile, and hence stored information is lost when the power supply is removed. This problem can be overcome by using a low consumption battery-maintained power supply operating in a standby mode, and hence if the normal power supply fails the battery supply automatically takes over.

A typical static RAM, which can be used as a *scratchpad* to store temporary results, variable data, etc., is the MCM6810, 128 × 8-bit memory.

In general, memory performance is characterised by three main parameters

(1) *Access time*, which refers to the time delay between the instant of application of the appropriate memory address and the instant of the corresponding memory output being available.

(2) *Writing time*, which refers to the time delay between the instant the write signal is enabled and the instant that information is deposited (stored) in the memory.

(3) *Cycle time*, which refers to the minimum time required for the memory to settle following a read/write operation before another read/write can be undertaken.

In using memory (RAM or ROM) addressing normally occurs at two levels, namely at the chip selection level and then at the location selection level. Most memory devices have a number (4-6) of chip select inputs, some may be active high and some active low. High-order address lines are normally used as chip select inputs such that their active levels select one chip only. The low-order address lines are normally used with each chip to select a particular location inside the selected chip. When a chip is not selected its data bus drivers exhibit a high impedance state, thereby isolating the chip from the bus.

A microprocessor system will typically use a ROM to store the program and a RAM as a scratchpad memory; these may be connected as shown in figure 5.44, where the RAM would use memory map addresses 0000–007F and the ROM would use memory map addresses F000–F3FF.

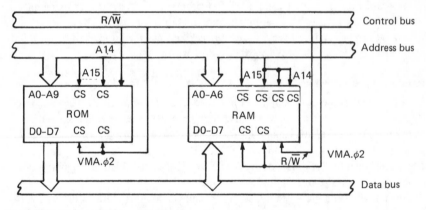

Figure 5.44 Typical ROM and RAM connections in a microprocessor system

Arithmetic Logic Unit (ALU)

An arithmetic logic unit: SN 74181, is capable of performing any one of 32 selectable operations, (16 arithmetic and 16 logic operations) on two 4-bit input words: A_0, A_1, A_2, A_3 and B_0, B_1, B_2 and B_3. A particular operation is selected using a 4-bit select word: S_0, S_1, S_2, S_3, and the mode input, M, see table 5.8. The ALU

Table 5.8 ALU functions

Select Word $S_3\,S_2\,S_1\,S_0$	Logic Function $M = 1$	Arithmetic Function $C_n = 1$	$C_n = 0$ $M = 0$
0 0 0 0	$F = \bar{A}$	$F = A$	$F = A$ Plus 1
0 0 0 1	$F = \overline{A + B}$	$F = A + B$	$F = (A + B)$ Plus 1
0 0 1 0	$F = \bar{A}B$	$F = A + \bar{B}$	$F = (A + \bar{B})$ Plus 1
0 0 1 1	$F = 0$	$F =$ Minus 1 (2's comp)	$F =$ Zero
0 1 0 0	$F = \overline{AB}$	$F = A$ Plus $A\,\bar{B}$	$F = A$ plus $A\,\bar{B}$ Plus 1
0 1 0 1	$F = \bar{B}$	$F = (A + B)$ Plus $A\,\bar{B}$	$F = (A + B)$ Plus $A\,\bar{B}$ Plus 1
0 1 1 0	$F = A \oplus B$	$F = A$ Minus B Minus 1	$F = A$ Minus B
0 1 1 1	$F = A\,\bar{B}$	$F = A\,\bar{B}$ Minus 1	$F = A\,\bar{B}$
1 0 0 0	$F = \bar{A} + B$	$F = A$ Plus AB	$F = A$ Plus AB Plus 1
1 0 0 1	$F = \overline{A \oplus B}$	$F = A$ Plus B	$F = A$ Plus B Plus 1
1 0 1 0	$F = B$	$F = (A + \bar{B})$ Plus AB	$F = (A + \bar{B})$ Plus AB Plus 1
1 0 1 1	$F = AB$	$F = AB$ Minus 1	$F = AB$
1 1 0 0	$F = 1$	$F = A$ Plus A†	$F = A$ Plus A Plus 1
1 1 0 1	$F = A + \bar{B}$	$F = (A + B)$ Plus A	$F = (A + B)$ Plus A Plus 1
1 1 1 0	$F = A + B$	$F = (A + \bar{B})$ Plus A	$F = (A + \bar{B})$ Plus A Plus 1
1 1 1 1	$F = A$	$F = A$ Minus 1	$F = A$

†Each bit is shifted to the next more significant position.

also accepts an input carry, C_n, and yields a 4-bit result: F_0, F_1, F_2, F_3, and an output carry C_{n+4}.

Several ALUs can be cascaded by connecting their carry inputs and outputs. With cascaded connections the carries take time to propagate when addition/subtraction operations are performed. However, look-ahead carry generators (SN74182) can be used with the cascaded ALUs to achieve fast operation.

APPENDIX 5.2 PROGRAM FOR EQUATION 5.18

```
S1130200000000101020203030404050506060707B4
S1130010080809090A0A030030C0C0C0D0D0E0E0F27
S11300200F1010111112121313141415151616179C
S1130030171818191A1A1B1B1C1C1D1D1E1E13
S11300421F1F2020212122222323242424252526 86
S11300502627272828292A2A232B2C2C2D2D2EFC
S11300602E2F2F3030303131323233333434353572
S11300703636373738383939393A3A3B3B3C3C3DE5
S11300803D3E3E3F3F4140404141424243434444 60
S11300904545464647474848494A4A4B4B4CD8
S11300A04C4D4D4E4E4F4F505051515252535354 4C
S11300305454555556565757585859595A5A5B5BC4
S11300C05C5C5D5D5E5E5F5F6061616262636364 30
S11300D06465656666676768686969696A6A6B6B6C 9C
S11300E06C6C6D6D6E6E6F6F7070717172727373 14
S11300F0747475757676777778787879797A7A 02
S1130100010202030405050607080909 0A0B0C0C 81
S113011000D0E0F1010111213141415161717181 9A9
S11301201A1B1B1C1D1E1E1F2021222223242525 D1
S11301302627282929 2A2B2C2C2D2E2F303132333 F7
S11301403334353637373839393A3B3B3C3D3E3E3F 16
S113015040414242434445454647484949 4A4B4C3D
S11301604C4D4E4F5050515253535455565757586 7
S1130170595A5A5B5C5D5E5E5F606162626364658E
S113018065666767686969 6A6B6C6C6D6E6F707071B7
S11301907273737475767777787979 7A7A7B7C7D7E DF
S11301A07E7F80 81318283848586878889898A08
S11301B08B8C8C8D8E8F90909192939394959697 2F
S11301C0979899 9A9A9B9C9D9E9E9FA0A1A1A2A359
S11301D0A4A5A5A6A7A8A8A9AAABACACADAEAFB0 30
S11301E0B0B1B2B3B3B4B5B6B7B7B839BABABBBCA9
S11301F0BDBEBEBFC0C1C1C2C3C4C5C5C6C7C7C7 D3
S113020000A2D0F4F5F0CB70200B78009B7800BB7F8
S113021080843B7800A86043B7800B863CB800900
S11302208634B78009863DB78009 0E3E0F7E021B D7
S1130230CE0100B68008F60200B70240F702429 6EB
S113024003E60A1B240236FFB7800AB702000C4F9C
S1050225055F3B0E
```


*   Address   Op-code field   Check sum

      field   field

APPENDIX 5.3 M6802 INSTRUCTION SET

(by kind permission of Motorola Semiconductors)

ACCUMULATOR AND MEMORY OPERATIONS	MNEMONIC	IMMED OP	IMMED ~	IMMED #	DIRECT OP	DIRECT ~	DIRECT #	INDEX OP	INDEX ~	INDEX #	EXTND OP	EXTND ~	EXTND #	INHER OP	INHER ~	INHER #	BOOLEAN/ARITHMETIC OPERATION (All register labels refer to contents)	H (5)	I (4)	N (3)	Z (2)	V (1)	C (0)
Add	ADDA	8B	2	2	9B	3	2	AB	5	2	BB	4	3				$A + M \rightarrow A$	↕	•	↕	↕	↕	↕
	ADDB	CB	2	2	DB	3	2	EB	5	2	FB	4	3				$B + M \rightarrow B$	↕	•	↕	↕	↕	↕
Add Acmltrs	ABA													1B	2	1	$A + B \rightarrow A$	↕	•	↕	↕	↕	↕
Add with Carry	ADCA	89	2	2	99	3	2	A9	5	2	B9	4	3				$A + M + C \rightarrow A$	↕	•	↕	↕	↕	↕
	ADCB	C9	2	2	D9	3	2	E9	5	2	F9	4	3				$B + M + C \rightarrow B$	↕	•	↕	↕	↕	↕
And	ANDA	84	2	2	94	3	2	A4	5	2	B4	4	3				$A \bullet M \rightarrow A$	•	•	↕	↕	R	•
	ANDB	C4	2	2	D4	3	2	E4	5	2	F4	4	3				$B \bullet M \rightarrow B$	•	•	↕	↕	R	•
Bit Test	BITA	85	2	2	95	3	2	A5	5	2	B5	4	3				$A \bullet M$	•	•	↕	↕	R	•
	BITB	C5	2	2	D5	3	2	E5	5	2	F5	4	3				$B \bullet M$	•	•	↕	↕	R	•
Clear	CLR							6F	7	2	7F	6	3				$00 \rightarrow M$	•	•	R	S	R	R
	CLRA													4F	2	1	$00 \rightarrow A$	•	•	R	S	R	R
	CLRB													5F	2	1	$00 \rightarrow B$	•	•	R	S	R	R
Compare	CMPA	81	2	2	91	3	2	A1	5	2	B1	4	3				$A - M$	•	•	↕	↕	↕	↕
	CMPB	C1	2	2	D1	3	2	E1	5	2	F1	4	3				$B - M$	•	•	↕	↕	↕	↕
Compare Acmltrs	CBA													11	2	1	$A - B$	•	•	↕	↕	↕	↕
Complement, 1's	COM							63	7	2	73	6	3				$\bar{M} \rightarrow M$	•	•	↕	↕	R	S
	COMA													43	2	1	$\bar{A} \rightarrow A$	•	•	↕	↕	R	S
	COMB													53	2	1	$\bar{B} \rightarrow B$	•	•	↕	↕	R	S
Complement, 2's (Negate)	NEG							60	7	2	70	6	3				$00 - M \rightarrow M$	•	•	↕	↕	①	②
	NEGA													40	2	1	$00 - A \rightarrow A$	•	•	↕	↕	①	②
	NEGB													50	2	1	$00 - B \rightarrow B$	•	•	↕	↕	①	②
Decimal Adjust, A	DAA													19	2	1	Converts Binary Add. of BCD Characters into BCD Format	•	•	↕	↕	③	③
Decrement	DEC							6A	7	2	7A	6	3				$M - 1 \rightarrow M$	•	•	↕	↕	④	•
	DECA													4A	2	1	$A - 1 \rightarrow A$	•	•	↕	↕	④	•
	DECB													5A	2	1	$B - 1 \rightarrow B$	•	•	↕	↕	④	•
Exclusive OR	EORA	88	2	2	98	3	2	A8	5	2	B8	4	3				$A \oplus M \rightarrow A$	•	•	↕	↕	R	•
	EORB	C8	2	2	D8	3	2	E8	5	2	F8	4	3				$B \oplus M \rightarrow B$	•	•	↕	↕	R	•
Increment	INC							6C	7	2	7C	6	3				$M + 1 \rightarrow M$	•	•	↕	↕	⑤	•
	INCA													4C	2	1	$A + 1 \rightarrow A$	•	•	↕	↕	⑤	•
	INCB													5C	2	1	$B + 1 \rightarrow B$	•	•	↕	↕	⑤	•

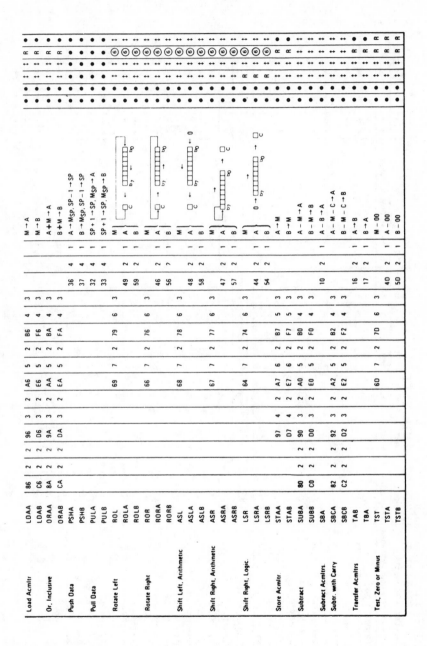

APPENDIX 5.3 continued

INDEX REGISTER AND STACK POINTER OPERATIONS

POINTER OPERATIONS	MNEMONIC	IMMED OP	IMMED ~	IMMED #	DIRECT OP	DIRECT ~	DIRECT #	INDEX OP	INDEX ~	INDEX #	EXTND OP	EXTND ~	EXTND #	INHER OP	INHER ~	INHER #	BOOLEAN/ARITHMETIC OPERATION	5 H	4 I	3 N	2 Z	1 V	0 C
Compare Index Reg	CPX	8C	3	3	9C	4	2	AC	6	2	BC	5	3				$(X_H/X_L) - (M/M + 1)$	•	•	(7)	↕	(8)	•
Decrement Index Reg	DEX													09	4	1	$X - 1 \to X$	•	•	•	↕	•	•
Decrement Stack Pntr	DES													34	4	1	$SP - 1 \to SP$	•	•	•	•	•	•
Increment Index Reg	INX													08	4	1	$X + 1 \to X$	•	•	•	↕	•	•
Increment Stack Pntr	INS													31	4	1	$SP + 1 \to SP$	•	•	•	•	•	•
Load Index Reg	LDX	CE	3	3	DE	4	2	EE	6	2	FE	5	3				$M \to X_H, (M + 1) \to X_L$	•	•	(9)	↕	R	•
Load Stack Pntr	LDS	8E	3	3	9E	4	2	AE	6	2	BE	5	3				$M \to SP_H, (M + 1) \to SP_L$	•	•	(9)	↕	R	•
Store Index Reg	STX				DF	5	2	EF	7	2	FF	6	3				$X_H \to M, X_L \to (M + 1)$	•	•	(9)	↕	R	•
Store Stack Pntr	STS				9F	5	2	AF	7	2	BF	6	3				$SP_H \to M, SP_L \to (M + 1)$	•	•	(9)	↕	R	•
Indx Reg → Stack Pntr	TXS													35	4	1	$X - 1 \to SP$	•	•	•	•	•	•
Stack Pntr → Indx Reg	TSX													30	4	1	$SP + 1 \to X$	•	•	•	•	•	•

JUMP AND BRANCH

OPERATIONS	MNEMONIC	RELATIVE OP	RELATIVE ~	RELATIVE #	INDEX OP	INDEX ~	INDEX #	EXTND OP	EXTND ~	EXTND #	INHER OP	INHER ~	INHER #	BRANCH TEST	5 H	4 I	3 N	2 Z	1 V	0 C
Branch Always	BRA	20	4	2										None	•	•	•	•	•	•
Branch If Carry Clear	BCC	24	4	2										$C = 0$	•	•	•	•	•	•
Branch If Carry Set	BCS	25	4	2										$C = 1$	•	•	•	•	•	•
Branch If = Zero	BEQ	27	4	2										$Z = 1$	•	•	•	•	•	•
Branch If ≥ Zero	BGE	2C	4	2										$N \div V = 0$	•	•	•	•	•	•
Branch If > Zero	BGT	2E	4	2										$Z + (N \div V) = 0$	•	•	•	•	•	•
Branch If Higher	BHI	22	4	2										$C + Z = 0$	•	•	•	•	•	•
Branch If ≤ Zero	BLE	2F	4	2										$Z + (N \div V) = 1$	•	•	•	•	•	•
Branch If Lower Or Same	BLS	23	4	2										$C + Z = 1$	•	•	•	•	•	•
Branch If < Zero	BLT	2D	4	2										$N \div V = 1$	•	•	•	•	•	•
Branch If Minus	BMI	2B	4	2										$N = 1$	•	•	•	•	•	•
Branch If Not Equal Zero	BNE	26	4	2										$Z = 0$	•	•	•	•	•	•
Branch If Overflow Clear	BVC	28	4	2										$V = 0$	•	•	•	•	•	•
Branch If Overflow Set	BVS	29	4	2										$V = 1$	•	•	•	•	•	•
Branch If Plus	BPL	2A	4	2										$N = 0$	•	•	•	•	•	•
Branch To Subroutine	BSR	8D	8	2											•	•	•	•	•	•
Jump	JMP				6E	4	2	7E	3	3				See Special Operations	•	•	•	•	•	•
Jump To Subroutine	JSR				AD	8	2	BD	9	3					•	•	•	•	•	•
No Operation	NOP										01	2	1	Advances Prog Cntr Only	•	•	•	•	•	•
Return From Interrupt	RTI										3B	10	1		(10)					
Return From Subroutine	RTS										39	5	1	See special Operations	•	•	•	•	•	•
Software Interrupt	SWI										3F	12	1		•	S	•	•	•	•
Wait for Interrupt	WAI										3E	9	1		•	(11)	•	•	•	•

CONDITIONS CODE REGISTER

OPERATIONS	MNEMONIC	INHER OP	INHER ~	INHER =	BOOLEAN OPERATION	5 H	4 I	3 N	2 Z	1 V	0 C
Clear Carry	CLC	0C	2	1	0 → C	•	•	•	•	•	R
Clear Interrupt Mask	CLI	0E	2	1	0 → I	•	R	•	•	•	•
Clear Overflow	CLV	0A	2	1	0 → V	•	•	•	•	R	•
Set Carry	SEC	0D	2	1	1 → C	•	•	•	•	•	S
Set Interrupt Mask	SEI	0F	2	1	1 → I	•	S	•	•	•	•
Set Overflow	SEV	0B	2	1	1 → V	•	•	•	•	S	•
Acmltr A → CCR	TAP	06	2	1	A → CCR	⑫					
CCR → Acmltr A	TPA	07	2	1	CCR → A	•	•	•	•	•	•

LEGEND:

OP   Operation Code (Hexadecimal);
~   Number of MPU Cycles;
=   Number of Program Bytes;
+   Arithmetic Plus;
–   Arithmetic Minus;
•   Boolean AND;
Msp   Contents of memory location pointed to be Stack Pointer;
+   Boolean Inclusive OR;
⊕   Boolean Exclusive OR;
$\overline{M}$ Complement of M;
↑   Transfer Into;
0   Bit = Zero;
00   Byte = Zero;
H   Half carry from bit 3;
I   Interrupt mask
N   Negative (sign bit)
Z   Zero (byte)
V   Overflow, 2's complement
C   Carry from bit 7
R   Reset Always
S   Set Always
↕   Test and set if true, cleared otherwise
•   Not Affected
CCR   Condition Code Register
LS   Least Significant
MS   Most Significant

CONDITION CODE REGISTER NOTES:

(Bit set if test is true and cleared otherwise)

1. (Bit V) Test: Result = 10000000?
2. (Bit C) Test: Result = 00000000?
3. (Bit C) Test: Decimal value of most significant BCD Character greater than nine? (Not cleared if previously set.)
4. (Bit V) Test: Operand = 10000000 prior to execution?
5. (Bit V) Test: Operand = 01111111 prior to execution?
6. (Bit V) Test: Set equal to result of N ⊕ C after shift has occurred.
7. (Bit N) Test: Sign bit of most significant (MS) byte of result = 1?
8. (Bit V) Test: 2's complement overflow from subtraction of LS bytes?
9. (Bit N) Test: Result less than zero? (Bit 15 = 1)
10. (All) Load Condition Code Register from Stack. (See Special Operations)
11. (Bit I) Set when interrupt occurs. If previously set, a Non-Maskable Interrupt is required to exit the wait state.
12. (ALL) Set according to the contents of Accumulator A.

APPENDIX 5.4
(courtesy of Texas Instruments)

TABLE 1. INSTRUCTION SYMBOLS

SYMBOL	MEANING
ACC	Accumulator
D	Data memory address field
I	Addressing mode bit
K	Immediate operand field
PA	3-bit port address field
R	1-bit operand field specifying auxiliary register
S	4-bit left-shift code
X	3-bit accumulator left-shift field

TABLE 2. TMS320C10 INSTRUCTION SET SUMMARY

ACCUMULATOR INSTRUCTIONS

MNEMONIC	DESCRIPTION	NO. CYCLES	NO. WORDS	OPCODE INSTRUCTION REGISTER 15 14 13 12 11 10 9 8 7 6 5 4 3 2 1 0
ABS	Absolute value of accumulator	1	1	0 1 1 1 1 1 1 1 1 0 0 0 1 0 0 0
ADD	Add to accumulator with shift	1	1	0 0 0 0 ◄—S► I ◄——D——►
ADDH	Add to high-order accumulator bits	1	1	0 1 1 0 0 0 0 0 I ◄——D——►
ADDS	Add to accumulator with no sign extension	1	1	0 1 1 0 0 0 0 1 I ◄——D——►
AND	AND with accumulator	1	1	0 1 1 1 1 0 0 1 I ◄——D——►
LAC	Load accumulator with shift	1	1	0 0 1 0 ◄—S► I ◄——D——►
LACK	Load accumulator immediate	1	1	0 1 1 1 1 1 1 0 ◄——K——►
OR	OR with accumulator	1	1	0 1 1 1 1 0 1 0 I ◄——D——►
SACH	Store high-order accumulator bits with shift	1	1	0 1 0 1 1 ◄X► I ◄——D——►
SACL	Store low-order accumulator bits	1	1	0 1 0 1 0 0 0 0 I ◄——D——►
SUB	Subtract from accumulator with shift	1	1	0 0 0 1 ◄—S► I ◄——D——►
SUBC	Conditional subtract (for divide)	1	1	0 1 1 0 0 1 0 0 I ◄——D——►
SUBH	Subtract from high-order accumulator bits	1	1	0 1 1 0 0 0 1 0 I ◄——D——►
SUBS	Subtract from accumulator with no sign extension	1	1	0 1 1 0 0 0 1 1 I ◄——D——►
XOR	Exclusive OR with accumulator	1	1	0 1 1 1 1 0 0 0 I ◄——D——►
ZAC	Zero accumulator	1	1	0 1 1 1 1 1 1 1 1 0 0 0 1 0 0 1
ZALH	Zero accumulator and load high-order bits	1	1	0 1 1 0 0 1 0 1 1 ◄——D——►
ZALS	Zero accumulator and load low-order bits with no sign extension	1	1	0 1 1 0 0 1 1 0 1 ◄——D——►

AUXILIARY REGISTER AND DATA PAGE POINTER INSTRUCTIONS

MNEMONIC	DESCRIPTION	NO. CYCLES	NO. WORDS	OPCODE INSTRUCTION REGISTER 15 14 13 12 11 10 9 8 7 6 5 4 3 2 1 0
LAR	Load auxiliary register	1	1	0 0 1 1 1 0 0 R I ◄——D——►
LARK	Load auxiliary register immediate	1	1	0 1 1 1 0 0 0 R ◄——K——►
LARP	Load auxiliary register pointer immediate	1	1	0 1 1 0 1 0 0 0 1 0 0 0 0 0 0 K
LDP	Load data memory page pointer	1	1	0 1 1 0 1 1 1 1 I ◄——D——►
LDPK	Load data memory page pointer immediate	1	1	0 1 1 0 1 1 1 0 0 0 0 0 0 0 0 K
MAR	Modify auxiliary register and pointer	1	1	0 1 1 0 1 0 0 0 I ◄——D——►
SAR	Store auxiliary register	1	1	0 0 1 1 0 0 0 R I ◄——D——►

TABLE 2. TMS320C10 INSTRUCTION SET SUMMARY (CONTINUED)

BRANCH INSTRUCTIONS

MNEMONIC	DESCRIPTION	NO. CYCLES	NO. WORDS	OPCODE / INSTRUCTION REGISTER (15 14 13 12 11 10 9 8 7 6 5 4 3 2 1 0)
B	Branch unconditionally	2	2	1 1 1 1 1 0 0 1 0 0 0 0 0 0 0 0 0 0 0 0 ←— BRANCH ADDRESS —→
BANZ	Branch on auxiliary register not zero	2	2	1 1 1 1 0 1 0 0 0 0 0 0 0 0 0 0 0 0 0 0 ←— BRANCH ADDRESS —→
BGEZ	Branch if accumulator ≥ 0	2	2	1 1 1 1 1 1 0 1 0 0 0 0 0 0 0 0 0 0 0 0 ←— BRANCH ADDRESS —→
BGZ	Branch if accumulator > 0	2	2	1 1 1 1 1 1 0 0 0 0 0 0 0 0 0 0 0 0 0 0 ←— BRANCH ADDRESS —→
BIOZ	Branch on $\overline{BIO}$ = 0	2	2	1 1 1 1 0 1 1 0 0 0 0 0 0 0 0 0 0 0 0 0 ←— BRANCH ADDRESS —→
BLEZ	Branch if accumulator ≤ 0	2	2	1 1 1 1 1 0 1 1 0 0 0 0 0 0 0 0 0 0 0 0 ←— BRANCH ADDRESS —→
BLZ	Branch if accumulator < 0	2	2	1 1 1 1 1 0 1 0 0 0 0 0 0 0 0 0 0 0 0 0 ←— BRANCH ADDRESS —→
BNZ	Branch if accumulator ≠ 0	2	2	1 1 1 1 1 1 1 0 0 0 0 0 0 0 0 0 0 0 0 0 ←— BRANCH ADDRESS —→
BV	Branch on overflow	2	2	1 1 1 1 0 1 0 1 0 0 0 0 0 0 0 0 0 0 0 0 ←— BRANCH ADDRESS —→
BZ	Branch if accumulator = 0	2	2	1 1 1 1 1 1 1 1 0 0 0 0 0 0 0 0 0 0 0 0 ←— BRANCH ADDRESS —→
CALA	Call subroutine from accumulator	2	1	0 1 1 1 1 1 1 1 1 0 0 0 1 1 0 0
CALL	Call subroutine immediately	2	2	1 1 1 1 1 0 0 0 0 0 0 0 0 0 0 0 0 0 0 0 ←— BRANCH ADDRESS —→
RET	Return from subroutine or interrupt routine	2	1	0 1 1 1 1 1 1 1 1 0 0 0 1 1 0 1

T REGISTER, P REGISTER, AND MULTIPLY INSTRUCTIONS

MNEMONIC	DESCRIPTION	NO. CYCLES	NO. WORDS	OPCODE / INSTRUCTION REGISTER (15 14 13 12 11 10 9 8 7 6 5 4 3 2 1 0)
APAC	Add P register to accumulator	1	1	0 1 1 1 1 1 1 1 1 0 0 0 1 1 1 1
LT	Load T register	1	1	0 1 1 0 1 0 1 0 1 ←——D——→
LTA	LTA combines LT and APAC into one instruction	1	1	0 1 1 0 1 1 0 0 1 ←——D——→
LTD	LTD combines LT, APAC, and DMOV into one instruction	1	1	0 1 1 0 1 0 1 1 1 ←——D——→
MPY	Multiply with T register, store product in P register	1	1	0 1 1 0 1 1 0 1 1 ←——D——→
MPYK	Multiply T register with immediate operand; store product in P register	1	1	1 0 0 ←————————K————————→
PAC	Load accumulator from P register	1	1	0 1 1 1 1 1 1 1 1 0 0 0 1 1 1 0
SPAC	Subtract P register from accumulator	1	1	0 1 1 1 1 1 1 1 1 0 0 1 0 0 0 0

TABLE 2. TMS320C10 INSTRUCTION SET SUMMARY (CONCLUDED)

				OPCODE															
		NO.	NO.	INSTRUCTION REGISTER															
MNEMONIC	DESCRIPTION	CYCLES	WORDS	15	14	13	12	11	10	9	8	7	6	5	4	3	2	1	0

CONTROL INSTRUCTIONS

MNEMONIC	DESCRIPTION	NO. CYCLES	NO. WORDS	OPCODE INSTRUCTION REGISTER
DINT	Disable interrupt	1	1	0 1 1 1 1 1 1 1 1 0 0 0 0 0 0 1
EINT	Enable interrupt	1	1	0 1 1 1 1 1 1 1 1 0 0 0 0 0 1 0
LST	Load status register	1	1	0 1 1 1 0 1 1 I ◄———D———►
NOP	No operation	1	1	0 1 1 1 1 1 1 1 1 0 0 0 0 0 0 0
POP	POP stack to accumulator	2	1	0 1 1 1 1 1 1 1 1 0 0 1 1 1 0 1
PUSH	PUSH stack from accumulator	2	1	0 1 1 1 1 1 1 1 1 0 0 1 1 1 0 0
ROVM	Reset overflow mode	1	1	0 1 1 1 1 1 1 1 1 0 0 0 1 0 1 0
SOVM	Set overflow mode	1	1	0 1 1 1 1 1 1 1 1 0 0 0 1 0 1 1
SST	Store status register	1	1	0 1 1 1 1 0 0 I ◄———D———►

I/O AND DATA MEMORY OPERATIONS

MNEMONIC	DESCRIPTION	NO. CYCLES	NO. WORDS	OPCODE INSTRUCTION REGISTER
DMOV	Copy contents of data memory location into next location	1	1	0 1 1 0 1 0 0 1 I ◄———D———►
IN	Input data from port	2	1	0 1 0 0 0 ◄PA► I ◄———D———►
OUT	Output data to port	2	1	0 1 0 0 1 ◄PA► I ◄———D———►
TBLR	Table read from program memory to data RAM	3	1	0 1 1 0 0 1 1 1 I ◄———D———►
TBLW	Table write from data RAM to program	3	1	0 1 1 1 1 1 0 1 I ◄———D———►

APPENDIX 5.5
(courtesy of Texas Instruments)

```
ALAWCOMP    320 FAMILY MACRO ASSEMBLER  2.1 83.076        08:29:32  10/26/8
                                                                   PAGE 000

0001                    IDT      'ALAWCOMP'
0002            ***
0003            *       'ALAWCOMP' PERFORMS AN A-LAW COMPRESSION.
0004            *        THE 13-BIT SIGN-MAGNITUDE INPUT X,
0005            *
0006            *       X = P X11 X10 ... X2 X1 X0
0007            *
0008            *       IS ENCODED AS AN 8-BIT SIGN-MAGNITUDE NUMBER Y,
0009            *
0010            *       Y = P S2 S1 S0 Q3 Q2 Q1 Q0 consisting of
0011            *
0012            *       POLARITY BIT: P,
0013            *       3-BIT SEGMENT NUMBER: S = S2 S1 S0
0014            *       4-BIT QUANTIZATION BIN NUMBER: Q = Q3 Q2 Q1 Q0
0015            *
0016            *       Y IS INVERTED BEFORE TRANSMISSION.
0017            *       PORT 0 IS USED FOR I/O.
0018            *
0019            *       WORST-CASE TIMING IN CYCLES: 14 INIT / 36 LOOP
0020            *       SPACE REQUIREMENTS IN WORDS: 11 DATA / 97 PROG
0021            *
0022            *       CONSTANTS:
0023            *
0024            *
0025    0001 ONE    EQU      1        =1
```

```
0026        0002   BIT4   EQU    2        = >0010 (ONE IN BIT 4)
0027        0003   MASK12 EQU    3        = >0FFF (12 ONES)
0028        0004   MASK8  EQU    4        = >00FF ( 8 ONES)
0029        0005   MASK4  EQU    5        = >000F ( 4 ONES)
0030        0006   MASK2  EQU    6        = >0003 ( 2 ONES)
0031               *
0032               *VARIABLES:
0033        0007   X      EQU    7        DATA INPUT (13 BITS)
0034        0008   Y      EQU    8        ENCODED DATA OUTPUT (8 BITS)
0035        0009   P      EQU    9        POLARITY OF DATA (0 FOR POS)
0036        000A   S      EQU    10       3-BIT SEGMENT NUMBER
0037        000B   Q      EQU    11       4-BIT QUANTIZATION BIN NUMBER
0038 0000
0039               *
0040               *
0041 0000          AORG   0
0042               *
0043 0000 7E01 INIT  LACK   1
0044 0001 5001       SACL   ONE
0045 0002 2401       LAC    ONE,4
0046 0003 5002       SACL   BIT4
0047 0004 2C01       LAC    ONE,12
0048 0005 1001       SUB    ONE
0049 0006 5003       SACL   MASK12
0050 0007 7EFF       LACK   >00FF
0051 0008 5004       SACL   MASK8
0052 0009 7E0F       LACK   >000F
0053 000A 5005       SACL   MASK4
0054 000B 7E03       LACK   >0003
0055 000C 5006       SACL   MASK2
0056               *
0057               * GET INPUT AND SAVE POLARITY
0058 000D 4007 START IN     X,0      INPUT DATA THRU PORT 0
0059 000E 2C01       LAC    ONE,12   POLARITY BIT MASK
0060 000F 7907       AND    X
0061 0010 5C09       SACH   P,4      0 FOR POS; 1 FOR NEG.
0062               *
0063               * STRIP TO LOW 12 BITS
0064 0011 2007       LAC    X
0065 0012 7903       AND    MASK12
0066 0013 5007       SACL   X
0067               *
0068               * S =BP -4 WHERE BP = BIT POSITION OF THE LEFTMOST '1'
0069               * IN X. TIND THE '1' THROUGH A BINARY SEARCH OF 3 MSB'S
0070               * OF X. STORE S AND LOAD X SHIFTED LEFT BY 16-S SO
0071               * THAT THE '1' AND FOUR FOLLOWING BITS ARE IN THE HIGH
0072               * HALF OF THE ACCUMULATOR. SEARCH BITS 4 THRU 11.
0073               *
0074 0014 2805 LMOST LAC    MASK4,8
0075 0015 7907       AND    X
0076 0016 FF00       BZ     EEE
     0017 0038
0077               *
0078 0018 2A06       LAC    MASK2,10    1100 0000
0079 0019 7907       AND    X
0080 001A FF00       BZ     CC
     001B 002A
0081               *
0082 001C 2B01       LAC    ONE,11      1000 0000
0083 001D 7907       AND    X
0084 001E FF00       BZ     B
     001F 0025
0085               *
0086 0020 7E07       LACK   7           1... ....
0087 0021 500A       SACL   S
0088 0022 2907       LAC    X,9
```

```
0089 0023 F900        B    XDONE
     0024 0058
0090                  *
0091 0025 7E06  B     LACK  6                    01.. ....
0092 0026 500A        SACL  S
0093 0027 2A07        LAC   X,10
0094 0028 F900        B     XDONE
     0029 0058
0095                  *
0096 002A 2901  CC    LAC   ONE,9              0010 0000
0097 002B 7907        AND   X
0098 002C FF00        BZ    D
     002D 0033
0099                  *
0100 002E 7E05        LACK  5                    001. ....
0101 002F 500A        SACL  S
0102 0030 2B07        LAC   X,11
0103 0031 F900        B     XDONE
     0032 0058
0104                  *
0105 0033 7E04  D     LACK  4                    0001 ....
0106 0034 500A        SACL  S
0107 0035 2C07        LAC   X,12
0108 0036 F900        B     XDONE
     0037 0058
0109                  *
0110 0038 2606  EEE   LAC   MASK2,6            0000 1100
0111 0039 7907        AND   X
0112 003A FF00        BZ    GG
     003B 004A
0113                  *
0114 003C 2701        LAC   ONE,7              0000 100
0115 003D 7907        AND   X
0116 003E FF00        BZ    F
     003F 0045
0117                  *
0118 0040 7E03        LACK  3                    0000 1...
0119 0041 500A        SACL  S
0120 0042 2D07        LAC   X,13
0121 0043 F900        B     XDONE
     0044 0058
0122                  *
0123 0045 7E02  F     LACK  2                    0000 01..
0124 0046 500A        SACL  S
0125 0047 2E07        LAC   X,14
0126 0048 F900        B     XDONE
     0049 0058
0127                  *
0128 004A 2501  GG    LAC   ONE,5              0000 0010
0129 004B 7907        AND   X
0130 004C FF00        BZ    SEGZ
     004D 0053
0131                  *
0132 004E 7E01        LACK  1                    0000 001.
0133 004F 500A        SACL  S
0134 0050 2F07        LAC   X,15
0135 0051 F900        B     XDONE
     0052 0058
0136                  * SEGMENT 0: _SSSQQQQ =X/2
0137 0053 2F07  SEGZ  LAC   X,15
0138 0054 5807        SACH  X
0139 0055 2007        LAC   X
0140 0056 F900        B     SIGN
     0057 005C
0141                  *
0142                  * REMOVE LEFTMOST '1' AND STORE Q
```

```
0143 0058 6202   XDONE   SUBH    BIT4
0144 0059 580B           SACH    Q
0145             *
0146             * FORM 8-BIT COMPRESSED WORS FIR Q, S, AND P.
0147 005A 200B           LAC     Q       Q:BITS 0-3 ____QQQQ

0148 005B 040A           ADD     S,4     S:BITS 4-6 _SSSQQQQ
0149 005C 0709   SIGN    ADD     P,7     PSSSQQQQ
0150             *
0151             * COMPLEMENT FOR TRANSMISSION AND OUTPUT
0152 005D 7804           XOR     MASK8
0153 005E 5008           SACL    Y
0154 005F 4808           OUT     Y,0     PORT 0
0155 0060 F900   FIN     B       FIN
     0061 0060
0156             *
0157                     END
NO ERRORS, NO WARNINGS
```

```
0001                         IDT     'ALAWEXP'
0002                  * * *
0003                  *      'ALAWEXP' PERFORM AN A-LAW EXPANSION. THE 8-BIT
0004                  *      DATA INPUT IS
0005                  *
0006                  *      Y = P S2 S1 S0 Q3 Q2 Q1 Q0    WHICH CONSISTS OF
0007                  *
0008                  *        POLARITY BIT: P
0009                  *        3-BIT SEGMENT NUMBER: S = S2 S1 S0
0010                  *        4-BIT QUANTIZATION NUMBER: Q = Q3 Q2 Q1 Q0
0011                  *
0012                  *      THE INPUT Y IS EXPANDED INTO A 13-BIT OUTPUT
0013                  *
0014                  *      X = P X11 X10 X9 ... X2 X1 X0 CONSISTING OF
0015                  *
0016                  *        POLARITY BIT:P
0017                  *        AND A 13-BIT MAGNITUDE (X12...X0)
0018                  *
0019                  *      PORT 0 IS USED FOR I/O.
0020                  *      WORST-CASE TIMING IN CYCLES: 4 INIT / 25 LOOP
0021                  *      SPACE REQUIREMENTS IN WORDS: 7 DATA / 48 LOOP
0022                  *
0023                  * CONSTANTS:
0024                  *
0025       0001  ONE    EQU     1        = 1
0026       0002  BIT7   EQU     2        = >0080 (ONE IN BIT 7)
0027                  *
0028                  * VARIABLES:
0029                  *
0030       0003  Y      EQU     3 .      A-LAW COMPRESSED 8-BIT DATA INPUT
0031       0004  X      EQU     4        DECODED (EXPANDED) 13-BIT OUTPUT
0032       0005  P      EQU     5        POLARITY OF DATA (0 FOR POS)
0033       0006  S      EQU     6        3-BIT SEGMENT NUMBER
0034       0007  SUM    EQU     7        VALUE TO BE SHIFTED
0035                  *
0036 0000              AORG    0
0037              *
0038 0000 7E01  INIT   LACK    1
0039 0001 5001         SACL    ONE
0040 0002 2701         LAC     ONE,7
0041 0003 5002         SACL    BIT7
0042              *
0043              * INVERT INPUT
0044              *
0045 0004 4003  START  IN      Y,0
0046 0005 7EFF         LACK    >00FF
0047 0006 7803         XOR     Y
0048 0007 5003         SACL    Y
0049              *
0050              * SAVE POLARITY AND STRIP TO LOW 7 BITS
0051 0008 7902         AND     BIT7
0052 0009 5005         SACL    P        0000 FOR POS; 0080 FOR NEG
0053 000A 7E7F         LACK    >007F
0054 000B 7903         AND     Y
0055 000C 5003         SACL    Y
0056              *
0057              * MAGNITUDE IS CORRECT. STRIP Y OF S AND Q.
0058 000D 2C03         LAC     Y,12     SHIFT S INTO HIGH HALF OF ACC
0059 000E 5806         SACH    S
0060 000F 2006         LAC     S        CHECK FOR SEGMENT 0
0061 0010 FE00         BNZ     SEGNZ
     0011 0016
0062              * SEGMENT 0: EXPAND X TO 2*Q + 1
0063 0012 2103         LAC     Y,1
0064 0013 0001         ADD     ONE
```

```
0065 0014 F900          B       SIGN
     0015 001D
0066                * NONZERO SEGMENT: SUM = 2*Q + 33
0067 0016 7E21  SEGNZ   LACK    33
0068 0017 0103          ADD     Y,1
0069 0018 1506          SUB     S,5      REMOVE S BITS
0070 0019 5007          SACL    SUM
0071            *
0072                * SHIFT SUM BY S USING VARIABLE SHIFT ROUTINE AT SBASE
0073 001A 7E20          LACK    SBASE-2 OFFSET (MINUS 0 CASE)
0074 001B 0106          ADD     S,1      DOUBLE S (2 WDS/SHIFT SEGMENT)
0075 001C 7F8C          CALA             SHIFT SUM BY S
0076            *
0077                * ACC = MAGNITUDE. ADD POLARITY TO BIT 12.
0078 001D 0505  SIGN    ADD     P,5      SHIFT P TO BIY 12
0079 001E 5004          SACL    X
0080 001F 4804          OUT     X,0      OUTPUT RESULT TO PORT 0
0081 0020 F900  FIN     B       FIN
     0021 0020
0082            *
0083                * LOAD SUM SHIFTED 0:6
0084 0022 2007  SBASE   LAC     SUM,0
0085 0023 7F8D          RET
0086 0024 2107          LAC     SUM,1
0087 0025 7F8D          RET
0088 0026 2207          LAC     SUM,2
0089 0027 7F8D          RET
0090 0028 2307          LAC     SUM,3
0091 0029 7F8D          RET
0092 002A 2407          LAC     SUM,4
0093 002B 7F8D          RET
0094 002C 2507          LAC     SUM,5
0095 002D 7F8D          RET
0096 002E 2607          LAC     SUM,6
0097 002F 7F8D          RET
0098            *
0099                    END
NO ERRORS, NO WARNINGS
```

APPENDIX 5.6 MANUFACTURERS AND SUPPLIERS OF DSP DEVICES

Advanced Micro Devices (UK) Ltd
AMD House, Goldsworth Road, Woking, Surrey GU21 1JJ, UK. Tel: (04862) 22121
29500 signal processor family (1982): a bit-slice architecture for DSP and array processing operations; benchmark: 1024-point complex FFT in approximately 2 ms

American Microsystems Inc.
3800 Homestead Road, Santa Clara, CA 95051, USA. Tel: (408) 246-0330
S28211 signal processor family (1982): architecture pipelined to perform a read, multiply and accumulate with each 300 ns instruction cycle

Analog Devices Ltd
Central Avenue, East Molesey, Surrey KT8 0SN, UK. Tel: 01-941 0466
ADSP-2100 (1986): a programmable single-chip microprocessor for DSP and
high-speed numeric processing applications; benchmark: 1024-point complex
FFT in 7.2 ms

Burr-Brown
PO Box 11400, Tucson, AZ 85734, USA. Tel: (602) 746-1111
MPV960 (1985): configured around the Burr-Brown ADC84 12-bit A/D
converter and the TMS320 DSP chip; provides high-speed analogue input with
DSP on a VMEbus board

Calmos Systems
20 Edgewater Street, Kanata, Ontario, Canada K2L 1V8. Tel: (613) 836-1014
CA128X16 FIR filter DMCU (1985): a building block for implementing FIR
filters of orders from 1 to 128 points; when used in conjunction with an external
filter coefficient memory and a multiplier-accumulator the data memory and
control unit (DMCU) provides an FIR filter implementation with four ICs

Euroka Oy
Hameenite 155 C52a, 00560 Helsinki, Finland. Tel: (358-0) 799522
SPU292 signal processing module: based on the Intel 2929 and designed to
provide a microcomputer system with analogue signal processing capability, that
is, for DSP applications

Fujitsu
*Amber Components Ltd, Rabans Close, Aylesbury, Bucks HP19 3RS, UK. Tel:
(0296) 34141*
MB8764 general-purpose CMOS digital signal processor (1985): features a high-
speed pipelined multiplier, supports concurrent operations with compound
instructions and multiple data paths, and offers flexible and expandable memory
options and an on-chip DMA channel; benchmark: 512-point FFT in 16 ms

General Instrument Microelectronics Ltd
Times House, Ruislip, Middx HA4 8LE, UK. Tel: (08956) 36141
DSP320C10-25 CMOS digital signal processor (1986): a 25 MHz device based on
a pipelined architecture, and capable of executing 6.25 MIPS

Gould AMI Semiconductors
AMI House, 56-58 Prospect Place, Swindon SN1 3JZ, UK. Tel: (0793) 37852
S7720 digital signal processor (1984): an advanced-architecture microcomputer
optimised for signal processing algorithms; benchmark: 64-point complex FFT
in 1.6 ms

Hitachi Electronic Components (UK) Ltd
21 Upton Road, Watford, Herts WD1 7TP, UK. Tel: (0923) 46488
HD61810B digital signal processor (1985): has a floating-point multiplier and a fixed-point/floating-point ALU, and can process signals over a dynamic range up to 32 bit, achieved via the automatic switching between fixed-point and floating-point arithmetic when data word values exceed 16 bit

Inmos Ltd
1000 Aztec West, Almondsbury, Bristol BS12 4SQ, UK. Tel: (0454) 616616
IMS A100 cascadable CMOS signal processor (1986): contains an array of 32 high-speed, high-accuracy, 16×16 bit multiplier-accumulators, plus registers and control logic, in a dataflow architecture to implement a digital transversal filter; benchmark: two A100s can achieve a 1024-point complex DFT in less than 2 ms

Intel Corp.
5000 W Williams Fields Road, Chandler, AZ 85224, USA. Tel: (602) 961-2000
2920 NMOS signal processor (1979): the first DSP chip, not now widely available

ITT Semiconductors
Intermetall, 145–147 Ewell Road, Surbiton, Surrey KT6 6AW, UK. Tel: 01–390 6577
UDP101 CMOS digital signal processor (1984): a mask-programmable device which has a multibus structure; uses pipelined program execution and the basic multiply and add instruction is implemented in 200 ms

Logic Devices Inc.
Abacus Electronics Ltd, Abacus House, Bone Lane, Newbury, Berks RG14 5SF, UK. Tel: (0635) 33311
LSH32 32-bit cascadable barrel shifter–normaliser (1986): a high-speed shifter for use in floating-point normalisation, word pack/unpack, field extraction and similard DSP applications

Loughborough Sound Images Ltd
The Technology Centre, Epinal Way, Loughborough, Leics LE11 0QE, UK. Tel: (0509) 231843
TMS320 board (1986): IBM PC plug-in for the development of DSP applications using the TMS32020 device

Motorola Ltd
Semiconductor Products Group, Colvilles Road, Kelvin Estate, East Kilbride, Glasgow G75 0TG, UK. Tel: (0734) 787848
DSP56000 family of DSP chips (1986): includes general-purpose, algorithm-

specific, application-specific and building-block parts; the DSP56000 device is capable of executing 10.25 MIPS and is suitable for general-purpose use; the DSP56200 device is a cascadable adaptive FIR filter, suitable for echo cancelling, telephone line equalisation and other DSP applications

Mullard Ltd
Mullard House, Torrington Place, London WC1E 7HD, UK. Tel: 01-580 6633
SP50 family of CMOS DSP devices (1986): the PCB5011 is a ROMless version of the PCB5010 DSP device with a highly parallel Harvard architecture; it has a 16 × 16 bit hardware multiplier aided by a barrel shifter and format adjuster; benchmark: 16-point complex FFT in under 30 μs (excluding initialisation procedures)

National Semiconductor (UK) Ltd
The Maple, Kembrey, Park, Swindon, Wilts SN2 6UT, UK. Tel: (0793) 614141
LM32900 digital signal processor (1986): CMOS microprocessor optimised for DSP operations; it uses a pipelined Harvard architecture and a multiply and add/subtract operation can be executed in 200 ns

NEC Electronics (UK) Ltd
Block 3, Carfin Industrial Estate, Motherwell ML1 4UL, UK. Tel: (0698) 732221
PD7720 digital signal processor (1980): uses a Harvard architecture and a multiply operation (16 × 16 bit) can be implemented in 250 ns; benchmark: FFT radix-2 butterfly in 9 μs

Plessey Semiconductors Ltd
Cheney Manor, Swindon, Wilts SN2 2QW, UK. Tel: (0793) 36251
MS2014FAD (1985), PDSP16401 (1986) and PDSP16112 (1986) digital signal processing chips: the MS2014FAD is a building block for implementing digital filters; the PDSP16401 is a two-dimensional edge detector; and the PDSP16112 is a 16 × 12 bit complex number multiplier

Pye TMC Ltd
Commercial Division, Martell Road, West Dulwich, London SE21 8EF, UK.
Tel: 01-670 2211
TMC539A PMOS dual second-order digital filter (*ca*. 1980): the filter sections are independent and may be used separately as a complete biquadratic, or they may be cascaded to form higher-order filters; sampling frequencies up to 16 kHz are practical

Reticon Corp.
910 Benicia Avenue, Sunnyvale, CA 94086, USA. Tel: (408) 738-4266
TAD32 tapped analogue delay (1977): a charge transfer device with 32 taps for implementing transversal filters; sampling rates up to 5 MHz are possible

Spectrum Devices Ltd
Central Avenue, East Molesey, Surrey KT8 0SN, UK. Tel: 01–941 2708
SDF family of digital filter building blocks (1981): a range of modules for
various digital filter applications; for example, a single-pole high-pass filter with
four selectable cut-off frequencies is provided by the SDF208S module, and it
may be operated with a sampling rate up to 15 MHz

STC Components Ltd
Semiconductors Division, Maidstone Road, Sidcup, Kent DA14 5HT. UK.
Tel: 01 300 3333
DSP128 digital signal processor (1984): a cascadable integrated signal processor
(CRISP) which may be connected on a common bus for efficient implementa-
tion of multiprocessing tasks; it may also be used as a standalone processor or be
interfaced to a general-purpose microprocessor; benchmark: 64-point complex
radix-2 FFT in 1.5 ms

Texas Instruments
Manton Lane, Bedford MK41 7PA, UK. Tel: (0234) 223000
TMS320 family of digital signal processors (1983): a combination of a general-
purpose single-chip microprocessor architecture with an ALU and multiplier,
resulting in a 32-bit internal Harvard architecture capable of 5 MIPS; benchmark
for the TMS320C25: 256-point complex FFT (radix-2 loop-coded) in 3.44 ms

Thomson Semiconducteurs
Ringway House, Bell Road, Daneshill, Basingstoke, Hants RG24 0QG, UK.
Tel: (0256) 29155
TS68930 programmable signal processing integrated VLSI (1986): a high-speed
general-purpose signal and arithmetic processor with a parallel/pipeline architecture to simultaneously execute one ALU function, multiplication, two reads and
one write operation and associated address calculation every 160 ns; benchmark:
256-point complex FFT in 2 ms

TRW Electronic Components Group
15-17 High Street, Bedford MK40 1RU, UK. Tel: (0234) 217711
TDC1028 and TMC2243 FIR Filters (1985): the TDC1028 can be clocked at a
rate up to 10 MHz and it implements an eight-tap FIR filter, whereas the
TMC2243 can be clocked at a faster rate of 20 MHz and it implements a three-
tap FIR filter

Answers to Problems

CHAPTER 1

1.1 $X(Z) = 2.5 - 1.2Z^{-1} - 0.08Z^{-2} + 8.9Z^{-3} + 0.4Z^{-4}$

1.2 (a) $\dfrac{Z}{Z - e^{-aT}} - \dfrac{Z}{Z - e^{-bT}}$; (b) $\dfrac{Z \sin \omega T}{Z^2 - 2Z \cos \omega T + 1}$

1.3 $f(z) = \dfrac{Z}{Z - \frac{1}{4}}$, $|Z| > \frac{1}{4}$

1.4 (a) $-12\delta(t) + 13(\frac{1}{4})^n$; (b) $2[1 - (0.5)^{n+1}]$

1.5 (a) $y(n)T = 1.2 - 0.2(-0.25)^n$
$y(0)T = 1.0$
$y(1)T = 1.25$
$y(2)T = 1.1875$
$y(3)T = 1.203125$
(b) $g(i)T = 2\delta(i)T - (-0.25)^i$

1.6 (a) Two zeros: at $Z = 0.25$ and $Z = -0.2$, two poles: at $Z = 0.4$ and $Z = -0.5$;
(b) Filter is stable.
(c) $g(i)T = 0.25\delta(i)T + 0.25(0.4)^i + 0.5(-0.5)^i$
(d) $y(n)T = [x(n)T - 0.05x(n-1)T - 0.05x(n-2)T - 0.1y(n-1)T + 0.2y(n-2)T]$
(e) $1.4286 \angle 0°$; (f) $1.3615 \angle 0°$

1.7 87.5%

CHAPTER 2

2.1 $G(Z) = \dfrac{1.571}{1 - 0.208Z^{-1}}$

2.2 (a) $\dfrac{0.481 - 0.962Z^{-1} + 0.481Z^{-2}}{1 - 0.672Z^{-1} + 0.253Z^{-2}}$

(b) $\dfrac{0.169 - 0.169Z^{-2}}{1 - 0.908Z^{-1} + 0.661Z^{-2}}$

(c) $\dfrac{0.830 - 0.908Z^{-1} + 0.830Z^{-2}}{1 - 0.98Z^{-1} + 0.661Z^{-2}}$

2.3 $G(Z) = \dfrac{1 - 0.819Z^{-1}}{1 - 1.549Z^{-1} + 0.779Z^{-2}}$

2.4 (i) $G(Z) = \dfrac{1 - 2Z^{-8} + Z^{-16}}{1 + 2Z^{-2} + Z^{-4}}$

(ii) $y(n)T = [x(n)T - 2x(n-8)T + x(n-16)T - 2y(n-2)T - y(n-4)T]$

(iii) $(1, 0, -2)$

2.5 $G(Z) = \dfrac{0.246 - 0.54Z^{-1} + 0.29Z^{-2}}{2.85 - 6.74Z^{-1} + 2.74Z^{-2}}$

2.6 $G(Z) = \dfrac{0.659Z(Z - 0.0437)}{(Z + 0.256 + j0.377)\,(Z + 0.256 - j0.377)}$

CHAPTER 3

3.1 $G(Z) = [-0.001 - 0.026Z^{-1} + 0.207Z^{-2} + 0.625Z^{-3} + 0.207Z^{-4} - 0.026Z^{-5} - 0.001Z^{-6}]$

3.2 $f_1(i) = a_0(i) + a_1(i) + a_2(i) - \delta_1 - 1$

$f_2(i) = a_0(i) + a_1(i) \cos \omega_1 T(i) + \frac{1}{2}\left\{(a_2(i)\,[1 + \cos 2\,\omega_1\,T(i)]\right\} + \delta_2$

$f_3(i) = a_1(i) + 2a_2(i) \cos \omega_1 T(i)$

$f_4(i) = a_0(i) - a_1(i) + a_2(i) - \delta_2$

3.3 $G(Z) = 0.25 - 1.5Z^{-1} + 0.25Z^{-2}$

3.4 $G(Z) = 0.2488 + 0.1034Z^{-1} + 0.05125Z^{-2} + 0.0295Z^{-3}$

CHAPTER 4

4.1 (a) $0.3639Q^2$

(b) $0.3100Q^2$

(c) $0.1860Q^2$

4.2 (a) 6

(b) 9

4.3 (a) $y(n)T = \pm 10$ for all n

(b) $y(n)T = \pm 1$ for all n

(c) $y(n)T = 0$ for all n

(d) $y(n)T = \pm 1$ for all n

dither signal $= -0.5(-1)^n$

CHAPTER 5

5.1 7

5.5 $y(n)T = -2x(n-1)T - x(n-2)T + x(n-4)T + 3x(n-5)T$; improvement factor $= 1.29$

Index